Introduction to Horticulture

Damien Miller

R CALLISTO REFERENCE

www.callistoreference.com

Callisto Reference,
118-35 Queens Blvd., Suite 400,
Forest Hills, NY 11375, USA

Visit us on the World Wide Web at:
www.callistoreference.com

ISBN: 978-1-64116-532-7 (Hardback)

Cataloging-in-Publication Data

Introduction to horticulture / Damien Miller.
    p. cm.
Includes bibliographical references and index.
ISBN 978-1-64116-532-7
1. Horticulture. 2. Agriculture. 3. Gardening. 4. Horticultural products.
5. Horticultural crops. I. Miller, Damien.
SB318 .I58 2022
635--dc23

Table of Contents

Preface

The agriculture of plants primarily for food, materials and beauty is termed as horticulture. It involves the cultivation and processing of flowers, fruits, vegetables and ornamental plants. It also includes plant propagation and cultivation to improve plant growth, yields, quality and nutrition value. Plants in horticulture can be grown for a variety of purposes such as food, non-food and social needs. Horticulture encompasses numerous fields such as plant conservation, soil management, landscape restoration, landscape design and arboriculture. Other major areas of horticulture are olericulture, pomology, viticulture, oenology and postharvest physiology. Olericulture deals with the production and marketing of vegetables. Pomology refers to the growing and selling of fruits. This textbook provides significant information of this discipline to help develop a good understanding of horticulture and its sub-fields. It is appropriate for students seeking detailed information in this area as well as for experts. Those in search of information to further their knowledge will be greatly assisted by this book.

A detailed account of the significant topics covered in this book is provided below:

Chapter 1- Horticulture is the cultivation of flowers, vegetables and fruits. Some of the main fields within horticulture are floriculture, landscape gardening, olericulture and arboriculture. It is also concerned with other processes such as plant conservation, landscape restoration and soil management. This is an introductory chapter which will introduce briefly all the major types of horticulture.

Chapter 2- Olericulture is the science and practice of growing vegetables for food. It also deals with the production, storage, processing and marketing of vegetables. There are many vegetable crops which fall under this category such as potato, tomato, carrot and onion. The diverse applications of olericulture in the current scenario have been thoroughly discussed in this chapter.

Chapter 3- The branch of botany that deals with the study and cultivation of fruits is termed as pomology. It is concerned with the enhancement, cultivation, development and physiological studies of fruit trees. It aims to enhance the fruit quality, reduce the cost of production and regulate the production period. This chapter has been carefully written to provide an easy understanding of pomology and the different fruit crops which are studied under it.

Chapter 4- Floriculture is a branch of horticulture that is concerned with the cultivation of ornamental and flowering plants for gardens and floral industry. It includes the cultivation of bedding plants, flowering pot plants and houseplants. Some of the major flower crops are rose, chrysanthemums, lilium and marigold. This chapter discusses in detail the processes and practices related to floriculture.

Chapter 5- The science and practice of growing fruits, vegetables, flowers and ornamental plants by following the essential principles of organic agriculture is termed as organic horticulture. It consists of a holistic and sustainable approach towards horticulture, often using natural processes which take place over an extended period of time. All the diverse principles of organic horticulture have been carefully analyzed in this chapter.

It gives me an immense pleasure to thank our entire team for their efforts. Finally in the end, I would like to thank my family and colleagues who have been a great source of inspiration and support.

Damien Miller

Chapter 1

Horticulture: An Introduction

Horticulture is the cultivation of flowers, vegetables and fruits. Some of the main fields within horticulture are floriculture, landscape gardening, olericulture and arboriculture. It is also concerned with other processes such as plant conservation, landscape restoration and soil management. This is an introductory chapter which will introduce briefly all the major types of horticulture.

Horticulture is the branch of plant agriculture dealing with garden crops, generally fruits, vegetables, and ornamental plants. The word is derived from the Latin hortus, "garden," and colere, "to cultivate." As a general term, it covers all forms of garden management, but in ordinary use it refers to intensive commercial production. In terms of scale, horticulture falls between domestic gardening and field agriculture, though all forms of cultivation naturally have close links.

Horticulture is divided into the cultivation of plants for food (pomology and olericulture) and plants for ornament (floriculture and landscape horticulture). Pomology deals with fruit and nut crops. Olericulture deals with herbaceous plants for the kitchen, including, for example, carrots (edible root), asparagus (edible stem), lettuce (edible leaf), cauliflower (edible flower), tomatoes (edible fruit), and peas (edible seed). Floriculture deals with the production of flowers and ornamental plants; generally, cut flowers, pot plants, and greenery. Landscape horticulture is a broad category that includes plants for the landscape, including lawn turf, but particularly nursery crops such as shrubs, trees, and climbers.

The specialization of the horticulturist and the success of the crop are influenced by many factors. Among these are climate, terrain, and other regional variations.

Horticultural Regions

Temperate Zones

Temperate zones for horticulture cannot be defined exactly by lines of latitude or longitude but are usually regarded as including those areas where frost in winter occurs, even though rarely. Thus most parts of Europe, North America, and northern Asia are included, though some parts of the United States, such as southern California and Florida, are considered subtropical. A few parts of the north coast of the Mediterranean and the Mediterranean islands are also subtropical. In the Southern Hemisphere, practically all of New Zealand, a few parts of Australia, and the southern part of South America have temperate climates. For horticultural purposes altitude is also a factor; the lower slopes of great mountain ranges, such as the Himalayas and the Andes, are included. Thus the temperate zones are very wide and the range of plants that can be grown in them is enormous, probably greater than in either the subtropical or tropical zones. In the temperate zones are the great coniferous and deciduous forests: pine, spruce, fir, most of the cypresses, the deciduous oaks (but excluding many of the evergreen ones), ash, birch, and linden (lime).

The temperate zones are also the areas of the grasses—the finest lawns particularly are in the regions of moderate or high rainfall—and of the great cereal crops. Rice is excluded as being tropical, but wheat, barley, corn (maize), and rye grow well in the temperate zones.

Plants in the temperate zones benefit from a winter resting season, which clearly differentiates them from tropical plants, which tend to grow continuously. Bulbs, annuals, herbaceous perennials, and deciduous trees become more frost-resistant with the fall of sap and therefore have a better chance of passing the resting season undamaged. Another influence is the varying length of darkness and light throughout the year, so that many plants, such as chrysanthemums, have a strong photoperiodism. The chrysanthemum flowers only in short daylight periods, although artificial lighting in nurseries can produce flowers the year round.

Most of the great gardens of the world have been developed in temperate zones. Particular features such as rose gardens, herbaceous borders, annual borders, woodland gardens, and rock gardens are also those of temperate-zone gardens. Nearly all depend for their success on the winter resting period.

Tropical Zones

There is no sharp line of demarcation between the tropics and the subtropics. Just as many tropical plants can be cultivated in the subtropics, so also many subtropical and even temperate plants can be grown satisfactorily in the tropics. Elevation is a determining factor. For example, the scarlet runner bean, a common plant in temperate regions, grows, flowers, and develops pods normally on the high slopes of Mt. Meru in Africa near the Equator; it will not, however, set pods in Hong Kong, a subtropical situation a little south of the Tropic of Cancer but at a low elevation.

In addition to elevation, another determinant is the annual distribution of rainfall. Plants that grow and flower in the monsoon areas, as in India, will not succeed where the climate is uniformly wet, as in Bougainville in the Solomon Islands. Another factor is the length of day, the number of hours the Sun is above the horizon; some plants flower only if the day is long, but others make their growth during the long days and flower when the day is short. Certain strains of the cosmos plant are so sensitive to light that, where the day is always about 12 hours, as near the Equator, they flower when only a few inches high; if grown near the Tropics of Cancer or Capricorn, they attain a height of several feet, if the seeds are sown in the spring, before flowering in the short days of autumn and winter. Poinsettia is a short-day plant that may be seen in flower in Singapore on any day of the year, while in Trinidad it is a blaze of glory only in late December.

In the tropics of Asia and parts of Central and South America the dominant features of the gardens are flowering trees, shrubs, and climbers. Herbaceous plants are relatively few, but many kinds of orchids can be grown.

Vegetable crops vary in kind and quality with the presence or absence of periodic dry seasons. In the uniformly wet tropics, the choice is limited to a few root crops and still fewer greens. Sweet potatoes grow and bear good crops where the average monthly rainfall, throughout the year, exceeds 10 inches (25 centimetres); they grow even better where there is a dry season. The same can be said of taro, yams, and cassava. Tropical greens from the Malay Peninsula are not as good as those grown in South China, the Hawaiian Islands, and Puerto Rico. They include several spinaches, of which Chinese spinach or amaranth is the best; several cabbages; Chinese onions and chives; and several gourds, cucumbers, and, where there is a dry season, watermelons. Brinjals, or eggplants, peppers, and okra are widely cultivated. Many kinds of beans can be grown successfully, including the French bean from the American subtropics, the many varieties of the African cowpea, and yard-long bean. The yam bean, a native of tropical America, is grown for its edible tuber. In the drier areas the pigeon pea, the soybean, the peanut (groundnut), and the Tientsin green bean are important crops. Miscellaneous crops include watercress, ginger, lotus, and bamboo.

Propagation

Propagation, the controlled perpetuation of plants, is the most basic of horticultural practices. Its two objectives are to achieve an increase in numbers and to preserve the essential characteristics of the plant. Propagation can be achieved sexually by seed or asexually by utilizing specialized vegetative structures of the plant (tubers and corms) or by employing such techniques as cutting, layering, grafting, and tissue culture.

Seed Propagation

The most common method of propagation for self-pollinated plants is by seed. In self-pollinated plants, the sperm nuclei in pollen produced by a flower fertilize egg cells of a flower on the same plant. Propagation by seed is also used widely for many cross-pollinated plants (those whose pollen is carried from one plant to another). Seed is usually the least expensive and often the only means of propagation and offers a convenient way to store plants over long periods of time. Seed kept dry and cool normally maintains its viability from harvest to the next planting season. Some can be stored for years under suitable conditions. Seed propagation also makes it possible to start plants free of most diseases. This is especially true with respect to virus diseases, because it is almost impossible to free plants of virus infections and because most virus diseases are not transmitted by seed. There are two disadvantages to seed propagation. First, genetic variation occurs in seed from cross-pollinated plants because they are heterozygous. This means that the plant grown from seed may not exactly duplicate the characteristics of its parents and may possess undesirable characteristics. Second, some plants take a long time to grow from seed to maturity. Potatoes, for example, do not breed true from seed and do not produce large tubers the first year. These disadvantages are overcome by vegetative propagation.

The practice of saving seed to plant the following year has developed into a specialized part of horticulture. Seed technology involves all of the steps necessary to ensure production of seed with high viability, freedom from disease, purity, and trueness to type. These processes may include specialized growing and harvesting techniques, cleaning, and distribution.

Relatively little tree and shrub seed is grown commercially; it is generally harvested from natural stands. Rootstock seed for fruit trees is often obtained as a by-product in fruit-processing industries. Seed growing and plant improvement are related activities. Thus many seed-producing firms actively engage in plant-breeding programs to accomplish genetic improvement of their material.

Harvesting of dry seed is accomplished by threshing. Seed from fleshy fruits is recovered through fermentation of the macerated (softened by soaking) pulp or directly from screening. Machines have been developed to separate and clean seed, based on size, specific gravity, and surface characteristics. Extended storage of seed requires low humidity and cool temperature.

Trade in seed requires quality control. For example, U.S. government seed laws require detailed labeling showing germination percentage, mechanical purity, amount of seed, origin, and moisture content. Seed testing is thus an important part of the seed industry.

While most vegetable seed germinates readily upon exposure to normally favourable environmental conditions, many seed plants that are vegetatively (asexually) propagated fail to germinate readily because of physical or physiologically imposed dormancy. Physical dormancy is due to structural limitations to germination such as hard impervious seed coats. Under natural conditions weathering for a number of years weakens the seed coat. Certain seeds, such as the sweet pea, have a tough husk that can be artificially worn or weakened to render the seed coat permeable to gases and water by a process known as scarification. This is accomplished by a number of methods including abrasive action, soaking in hot water, or acid treatment. Physiologically imposed dormancy involves the presence of germination inhibitors. Germination in such seed may be accomplished by treatment to remove these inhibitors. This may involve cold stratification, storing seed at high relative humidity and low temperatures, usually slightly above freezing. Cold stratification is a prerequisite to the uniform germination of many temperate-zone species such as apple, pear, and redbud.

Vegetative Propagation

Asexual or vegetative reproduction is based on the ability of plants to regenerate tissues and parts. In many plants vegetative propagation is a completely natural process; in others it is an artificial one. Vegetative propagation has many advantages. These include the unchanged perpetuation of naturally cross-pollinated or heterozygous plants and the possibility of propagating seedless progeny. This means that a superior plant may be reproduced endlessly without variation. In addition, vegetative propagation may be easier and faster than seed propagation, because seed dormancy problems are eliminated and the juvenile nonflowering stage of some seed-propagated plants is eliminated or reduced.

Vegetative propagation is accomplished by use of (1) apomictic seed; (2) specialized vegetative structures such as runners, bulbs, corms, rhizomes, offshoots, tubers, stems, and roots; (3) layers and cuttings; (4) grafting and budding; and (5) tissue culture.

Apomixis

Apomixis, the development of asexual seed (seed not formed via the normal sexual process), is a form of vegetative propagation for some horticultural plants including Kentucky bluegrass, mango, and citrus. Virus-free progeny can be produced in oranges from a seed that is formed from the nucellus, a maternal tissue.

Vegetative Structures

Many plants produce specialized vegetative structures that can be used in propagation. These may be storage organs such as tubers that enable the plant to survive adverse conditions or organs adapted for natural propagation—runners or rhizomes—so that the plant may rapidly spread.

Bulbs consist of a short stem base with one or more buds protected by fleshy leaves. They are found in such plants as the onion, daffodil, and hyacinth. Bulbs commonly grow at ground level, though bulblike structures (bulbils) may form on aerial stems in some lilies or in association with flower parts, as in the onion. Buds in the axils (angle between leaf and stem) of the fleshy leaves may form miniature bulbs (bulblets) that when grown to full size are known as offsets. Corms are short, fleshy, underground stems without fleshy leaves. The gladiolus and crocus are propagated by corms. They may produce new cormels from fleshy buds. Rhizomes are horizontal, underground stems that are compressed, as in the iris, or slender, as in turf grasses. Runners are specialized aerial stems, a natural agent of increase and spread for such plants as the strawberry, strawberry geranium, and bugleweed (Ajuga). Tubers are fleshy enlarged portions of underground stem. The edible portion of the potato, the tuber, is also used as a means of propagation.

A number of plants form lateral shoots from the stem, which when rooted serve to propagate the plant. These are known collectively as offshoots but are often called offsets, crown divisions, ratoons, or slips.

Roots may also be structurally modified as propagative and food-storage organs. These tuberous roots, fleshy swollen structures, readily form shoots (called adventitious, because they do not form from nodes). The sweet potato and dahlia are propagated by tuberous roots. Shoots that rise adventitiously from roots are called suckers. The red raspberry is propagated by suckers.

Layering and Cutting

Propagation can be accomplished by methods in which plants are induced to regenerate missing parts, usually adventitious roots or shoots. When the regenerated part is still attached to the plant the process is called layerage, or layering; When the regenerating portion is detached from the plant the process is called cuttage, or cutting.

Layering often occurs naturally. Drooping black raspberry stems tend to root in contact with the soil. The croton, a tropical plant, is commonly propagated by wrapping moist sphagnum enclosed in plastic around a stem cut to induce rooting. After rooting, the stem is detached and planted. Though simple and effective, layering is not normally adapted to large-scale nursery practices.

Cutting is one of the most important methods of propagation. Many plant parts can be used; thus cuttings are classified as root, stem, or leaf. Stem cuttings are the most common.

The ability of stems to regenerate missing parts is variable; consequently plants may be easy or difficult to root. The physiological ability of cuttings to form roots is due to an interaction of many factors. These include transportable substances in the plant itself: plant hormones (such as auxin), carbohydrates, nitrogenous substances, vitamins, and substances not yet identified. Environmental factors such as light, temperature, humidity, and oxygen are important, as are age, position, and type of stem.

Although easy-to-root plants such as willow or coleus can be propagated merely by plunging a stem in water or moist sand, the propagation of difficult-to-root species is a highly technical process. To achieve success with difficult-to-root plants special care is taken to control the environment and encourage rooting. A number of growth regulators stimulate rooting. A high degree of success has been achieved with indolebutyric acid, a synthetic auxin that is applied to the cut surface. A number of materials known as rooting cofactors have been found that interact with auxin to further stimulate rooting, and these are sold as a hormone rooting compound.

Humidity control is particularly important to prevent death of the stem from desiccation before rooting is complete. The use of an intermittent-mist system in propagation beds has proved to be an important means of improving success in propagation by cuttings. These operate by applying water to the plant for a few seconds each minute.

Grafting

Grafting involves the joining together of plant parts by means of tissue regeneration. The part of the combination that provides the root is called the stock; the added piece is called the scion. When more than two parts are involved, the middle piece is called the interstock. When the scion consists of a single bud, the process is called budding. Grafting and budding are the most widely used of the vegetative propagation methods.

Stock cambium and scion cambium respond to being cut by forming masses of cells (callus tissues) that grow over the injured surfaces of the wounds. The union resulting from interlocking of the callus tissues is the basis of graftage. In dicots (e.g., most trees) cambium—a layer of actively dividing cells between xylem (wood) and phloem (bast) tissues—is usually arranged in a continuous ring; in woody members, new layers of tissue are produced annually. Monocot stems (e.g., lilacs, orchids) do not possess a continuous cambium layer or increase in thickness; grafting is seldom possible.

The basic technique in grafting consists of placing cambial tissues of stock and scion in intimate association, so that the resulting callus tissue produced from stock and scion interlocks to form a living continuous connection. A snug fit can be obtained through the tension of the split stock and scion or both. Tape, rubber, and nails can be used to achieve close contact. In general, grafts are only compatible between the same or closely related species. Success in grafting depends on skill in achieving a snug fit. Warm temperatures (80°–85° F [27°–30° C]) increase callus formation and improve "take" in grafting. Thus grafts using dormant material are often stored in a warm, moist place to stimulate callus formation.

In grafting and budding, the rootstock can be grown from seed or propagated asexually. Within a year a small amount of scion material from one plant can produce hundreds of plants.

Grafting has uses in addition to propagation. The interaction of rootstocks may affect the performance of the stock through dwarfing or invigoration and in some cases may affect quality. Further, the use of more than one component can affect the disease resistance and hardiness of the combination.

Grafting as a means of growth control is used extensively with fruit trees and ornamentals such as roses and junipers. Fruit trees are normally composed of a scion grafted onto a rootstock. Sometimes an interstock is included between the scion and stock. The rootstock may be grown from seed (seedling rootstock) or asexually propagated (clonal rootstock). In the apple, a great many clonal rootstocks are available to give a complete range of dwarfing; rootstocks are also available to invigorate growth of the scion cultivar.

Tissue-culture techniques utilizing embryos, shoot tips, and callus can be used as a method of propagation. The procedure requires aseptic techniques and special media to supply inorganic elements; sugar; vitamins; and, depending on the tissue, growth regulators and organic complexes such as coconut milk, yeast, and amino-acid extract.

Embryo culture has been used to produce plants from embryos that would not normally develop within the fruit. This occurs in early-ripening peaches and in some hybridization between species. Embryo culture can also be used to circumvent seed dormancy.

A shoot tip, when excised and cultured, may produce roots at the base. This technique is employed for the purpose of producing plants free of disease. Certain orchids are rapidly multiplied by this method. Cultured shoot tips form an embryo-like stage that can be sectioned indefinitely to build up large stocks rapidly. These bulblike bodies left unsectioned develop into small plantlets. A similar procedure is used with the carnation, in which the shoot tip forms a cell mass that can be subdivided.

Callus-tissue culture—a very specialized technique that involves growth of the callus, followed by procedures to induce organ differentiation—has been successful with a number of plants including carrot, asparagus, and tobacco. Used extensively in research, callus culture has not been considered a practical method of propagation. Callus culture produces genetic variability because in some cases cells double their chromosome number. In rice and tobacco, mature plants have been obtained from callus formed from pollen. These plants have half the normal number of chromosomes.

Breeding

The isolation and production of superior types known as cultivars are the very keystones of horticulture. Plant breeding, the systematic improvement of plants through the application of genetic principles, has placed improvement of horticultural plants on a scientific basis. The raw material of improvement is found in the great variation that exists between cultivated plants and related wild species. The incorporation of these changes into cultivars adapted to specific geographical areas requires a knowledge of the theoretical basis of heredity and art and the skill to discover, perpetuate, and combine these small but fundamental differences in plant material.

The goal of the plant breeder is to create superior crop varieties. The cultivated variety, or cultivar, can be defined as a group of crop plants having similar but distinguishable characteristics. The term cultivar has various meanings, however, depending on the mode of reproduction of the crop. With reference to asexually propagated crops, the term cultivar means any particular clone considered of sufficient value to be graced with a name. With reference to sexually propagated crops, the concept of cultivar depends on the method of pollination. The cultivar in self-pollinated crops is basically a particular homozygous genotype, a pure line. In cross-pollinated crops the cultivar is not necessarily typified by any one plant but sometimes by a particular plant population, which at any one time is composed of genetically distinguishable individuals.

Environmental Control

Control of the natural environment is a major part of all forms of cultivation, whatever its scale. The scale, intensiveness, and economic risk in commercial gardening and nurseries, however, often require approaches markedly different from those of the small home garden.

The intensive cultivation practiced in horticulture relies on extensive control of the environment for all phases of plant life. The most basic environmental control is achieved by location and site: sunny or shady sites, proximity to bodies of water, altitude, and latitude.

Structures

Various structures are used for temperature control. Cold frames, used to start plants before the normal growing season, are low enclosed beds covered with a removable sash of glass or plastic. Radiant energy passes through the transparent top and warms the soil directly. Heat, however, as long-wave radiation, is prevented from leaving the glass or plastic cover at night. Thus heat that builds up in the cold frame during the day aids in warming the soil, which releases its heat gradually at night to warm the plants. When supplemental heat is provided, the structures are called hotbeds. At first, supplemental heat was supplied by respiration through the decomposition of manure or other organic matter. Today, heat is provided by electric cables, steam, or hot-water pipes buried in the soil.

Greenhouses are large hotbeds, and in most cases the source of heat is steam. While they were formerly made of glass, plastic films are now extensively used. Modern greenhouse ranges usually have automatic temperature control. Summer temperatures can be regulated by shading or evaporative "fan-and-pad" cooling devices. Air-conditioning units are usually too expensive except for scientific work. Greenhouses with precise environmental controls are known as phytotrons. Other environmental factors are controlled through automatic watering, regulation of light and shade, addition of carbon dioxide, and the regulation of fertility.

Shade houses are usually walk-in structures with shading provided by lath or screening. Summer propagation is often located in shade houses to reduce excessive water loss by transpiration.

Temperature Control

A number of temperature-control techniques are used in the field, including application of hot caps, cloches, plastic tunnels, and mulches of various types. Hot caps are cones of translucent paper or plastic that are placed over the tops of plants in the spring. These act as miniature greenhouses. In the past small glass sash called cloches were placed over rows to help keep them warm. Polyethylene tunnels supported by wire hoops that span the plants are now used for the same purpose. As spring advances the tunnels are slashed to prevent excessive heat buildup. In some cases the plastic tunnels are constructed so that they can be opened and closed when necessary. This technique is widely used in Israel for early production of vegetables.

Mulching is important in horticulture. Whether in the form of a topdressing of manure or compost or plastic sheeting, mulches offer the grower the various benefits of economical plant feeding, conservation of moisture, and control of weeds and erosion. Winter mulches are commonly used to protect such sensitive and valuable plants as strawberries and roses.

The storage of perishable plant products is accomplished largely through the regulation of their temperature to retard respiration and microbial activity. Excess water loss can be prevented by controlling humidity. Facilities that utilize the temperature of the atmosphere are called common storage. The most primitive types take advantage of the reduced temperature fluctuations of the soil by using caves or unheated cellars. Aboveground structures must be insulated and ventilated.

Complete temperature-regulated storages utilizing refrigeration and heating are now common for storage of horticultural products. The regulation of oxygen and carbon dioxide levels along with the regulation of temperature is known as controlled-atmosphere storage. Rooms are sealed so that gaseous exchange can be effectively controlled. Many horticultural products, such as fruit, can be kept fresh for as long as a year under these controlled conditions.

Frost Control

Frost is one of the high-risk elements for commercial growers, and the problem is accentuated by the fact that growers are striving to produce early-season crops. The precautions are consequently far more elaborate and costly than those of the domestic garden. Frost is especially damaging to perennial fruit crops in the spring—because flower parts are sensitive to freezing injury—and to tender transplants. The two weather conditions that produce freezing temperatures are rapid radiational cooling at night and introduction of a cold air mass with temperatures below freezing. Radiation frost occurs when the weather is clear and calm; air-mass freezes occur when it is overcast and windy.

Frost-control methods involve either reduction of radiational heat loss or conservation or addition of heat. Radiational heat loss may be reduced by hot caps, cold frames, or mulches. Heat may also be added from the air. Wind machines that stir up the air, for example, provide heat when temperature inversions trap cold air under a layer of warm air. These have been used extensively in citrus groves. Heat may be added directly by using heaters, usually fueled with oil. Sprinkler irrigation can also be used for frost control. The formation of ice is accompanied by the release of large amounts of heat, which maintains plants at the freezing temperature as long as the water is being frozen. Thus continuous sprinkling during frosty nights has been used to protect strawberries from frost injury.

Frost injury to transplants can be prevented through processes that increase the plant's ability to survive the impact of unfavourable environmental stress. This is known as hardening off. Hardening off of plants prior to transplanting can be accomplished by withholding water and fertilizer, especially nitrogen. This prevents formation of succulent tissue that is very frost-tender. Gradual exposure to cold is also effective for hardening. Induced cold resistance in crops such as cabbage, for example, can have a considerable effect; unhardened cabbages begin to show injury at 28° F (−2.2 °C), while hardened plants withstand temperatures as low as 22 °F (−5.6 °C).

Light Control

Light has a tremendous effect on plant growth. It provides energy for photosynthesis, the process by which plants, with the aid of the pigment chlorophyll, synthesize carbon compounds from water and carbon dioxide. Light also influences a great number of physiological reactions in plants. At energy values lower than those required for photosynthesis, light affects such processes as dormancy, flowering, tuberization, and seed-stalk development. In many cases these processes are affected by the length of day; the recurrent cycle of light is known as the photoperiod.

The control of light in horticultural practices involves increasing energy values for photosynthesis and controlling day length. Light is controlled in part by site and location. In the tropics day length approaches 12 hours throughout the year, whereas in polar regions it varies from zero to 24 hours. Light is also partly controlled by plant distribution and density.

Supplemental illumination in greenhouses increases photosynthesis. The cost of power to supply the artificial light, however, makes this impractical for all but crops of the highest value. Fluorescent lights are the most efficient for photosynthesis; special lights, rich in the wavelengths required, are now available.

Extension of day length through supplemental illumination and shading is common practice in the production of greenhouse flower crops, which are often induced to flower out of season. Artificial lengthening of short days, or interruption of the dark period, promotes flowering in long-day plants such as lettuce and spinach and prevents flowering of short-day plants such as chrysanthemums. Similarly, during naturally long days, shading to reduce day length prevents flowering of long-day plants and promotes flowering of short-day plants. The manipulation of day length is standard practice to control flowering of greenhouse chrysanthemums throughout the year. Tungsten lights have proved very effective for extending day length because they are rich in the red end of the spectrum that affects the photoperiodic reaction. Extending the day length is a relatively affordable practice because only a low light intensity is required. The same effects can be obtained through interruption of the dark period, even with light flashes. Decreasing day length is usually accomplished by simply covering the plants with black shade cloth.

Soil Management

The principles involved here are again similar to those of home gardening. But the financial considerations of horticulture naturally require a more scientific approach to soil care. To be successful, the grower must ensure the economic use of every square yard of ground, especially because the cost of sound horticultural land is among the highest of any in agriculture. Crop rotation is planned to ensure that the soil is not depleted of essential chemicals by repeated use of one type of plant in the same plot. Soil analysis is employed so that any such depletion can be rectified promptly. Fertilizers are applied in a precise routine and, of course, in a variety beyond the reach or needs of the ordinary gardener. They are frequently applied through leaves or stems in the form of chemical sprays.

Water Management

Depending on the terrain, water management may involve extensive works for irrigation and drainage. While the home gardener may well be content with a rough-and-ready appraisal of the wetness or dryness of the soil, horticulture is more exacting. Production of the high-quality fruits and vegetables demanded by the modern market requires a precise all-year balance of soil moisture, adjusted to the needs of the particular crop. These considerations apply whether the grower is situated in a high-rainfall area of Europe or in the parched land of the southwestern United States or Israel.

There are a number of general methods of land irrigation. In surface irrigation water is distributed over the surface of soil. Sprinkler irrigation is application of water under pressure as simulated rain. Subirrigation is the distribution of water to soil below the surface; it provides moisture to crops by upward capillary action. Trickle irrigation involves the slow release of water to each plant through small plastic tubes. This technique is adapted both to field and to greenhouse conditions.

Removal of excess water from soils can be achieved by surface or subsurface drainage. Surface drainage refers to the removal of surface water by development of the slope of the land utilizing systems of drains to carry away the surplus water. In subsurface drainage open ditches and tile fields intercept groundwater and carry it off. The water enters the tiling through the joints, and drainage is achieved by gravity feed through the tiles.

Pest Control

Horticultural plants are subject to a wide variety of injuries caused by other organisms. Plant pests include viruses, bacteria, fungi, higher plants, nematodes, insects, mites, birds, and rodents. Various methods are used to control them. The most successful treatments are preventive rather than curative.

Control of pests is achieved through practices that prevent harm to the plant and methods that affect the plant's ability to resist or tolerate intrusion by the pathogen. These can be classified as cultural, physical, chemical, or biological.

Traditional practices that reduce effective pest population include the elimination of diseased or infected plants or seeds (roguing), cutting out of infected plant parts (surgery), removal of plant debris that may harbour pests (sanitation), and alternating crops unacceptable to pests (rotation). Any of a number of techniques can be employed to render the environment unfavourable to the pest, such as draining or flooding and changing the soil's level of acidity or alkalinity.

Physical methods can be used to protect the plant against intrusion or to eliminate the pest entirely. Physical barriers range from the traditional garden fence to bags that protect each fruit, a common practice in Japan. Heat treatment is used to destroy some seed-borne pathogens and is a standard soil treatment in greenhouses to eliminate soil pests such as fungi, nematodes, and weed seed. Cultivation and tillage are standard practices for weed control.

The horticultural industry is now dependent upon chemical control of pests through pesticides, materials toxic to the pest in some stage of its life cycle. Commercial growers of practically all horticultural crops rely on complete schedules utilizing many different compounds. Pesticides are usually classed according to the organism they control: for example, bactericide, fungicide, nematicide, miticide, insecticide, rodenticide, and herbicide.

Selectivity of pesticides, the ability to discriminate between pests, is a relative concept. Some non-selective pesticides kill indiscriminately; most are selective to some degree. Most fungicides, for example, are not bactericidal. The development of highly selective herbicides makes it possible to destroy weeds from crops selectively. Selectivity can be achieved through control of dosage, timing, and method of application.

Plant pests can also be controlled through the manipulation of biological factors. This may be achieved through directing the natural competition between organisms or by incorporating natural resistance to the whole plant. The introduction of natural parasites or predators has been a successful method for the control of certain insects and weeds. Incorporation of genetic resistance is an ideal method of control. Thus breeding for disease and insect resistance is one of the chief goals of plant breeding programs. A major obstacle to this method of control is the ability of pathogens (disease-producing organisms) to mutate easily and attack previously resistant plants.

Growth Regulation by Chemicals

Control of plant growth through growth-regulating materials is a modern development in horticulture. These materials have resulted from basic investigations into growth and development, as well as systematic screening of materials to find those that affect differentiation and growth. This field was given great impetus by the discovery of a class of plant hormones known as auxins, which affect cell elongation.

Auxins have been correlated with inhibition and stimulation of growth as well as differentiation of organs and tissues. Such processes as cell enlargement, leaf and organ separation, budding, flowering, and fruit set (the formation of the fruit after pollination) and growth are influenced by auxins. In addition, auxins have been associated with the movement of plants in response to light and gravity. Auxin materials are used in horticulture for the promotion of rooting, fruit setting, fruit thinning, and fruit-drop control.

Gibberellins are a group of related, naturally occurring compounds of which only one, gibberellic acid, is commercially available. Gibberellins have many effects on plant development. The most startling is the stimulation of growth in many compact or dwarf plants. Minute applications transform bush to pole beans or dwarf to normal corn. Perhaps the most widespread horticultural use has been in grape production. The application of gibberellin is now a regular practice for the culture of the 'Thompson seedless' cultivar ("Sultanina") of grapes to increase berry size. In Japan applications of gibberellic acid are used to induce seedlessness in certain grapes.

Cytokinins are a group of chemical substances that have a decisive influence on the stimulation of cell division. In tissue culture high auxin and low cytokinin give rise to root development; low auxin and high cytokinin encourage shoot development.

Ethylene, a hydrocarbon compound, acts as a plant hormone to stimulate fruit ripening as well as rooting and flowering of some plants. An ethylene-releasing compound, 2-chloroethylphosphonic acid, has many horticultural applications, of which the most promising may be uniform ripening of tomatoes and the stimulation of latex flow in rubber.

Many compounds that inhibit growth hormones have application in horticulture. For example, a number of materials that inhibit formation of gibberellins by the plant cause dwarfing. These include chlorinated derivatives of quaternary ammonium and phosphonium compounds. Many of these have applications in floriculture. Growth retardants such as succinic acid–2,2-dimethyl-hydrazide, a gibberellin suppressor, have applications in horticulture from a wide array of effects that include dwarfing and fruit maturity. The growth inhibitor maleic hydrazide has been effective in preventing the sprouting of onions and potatoes.

Ornamental Horticulture

Ornamental horticulture consists of floriculture and landscape horticulture. Each is concerned with growing and marketing plants and with the associated activities of flower arrangement and landscape design. The turf industry is also considered a part of ornamental horticulture. Although flowering bulbs and flower seed represent an important component of agricultural production for the Low Countries of Europe, ornamentals are relatively insignificant in world trade.

Floriculture has long been an important part of horticulture, especially in Europe and Japan, and accounts for about half of the nonfood horticultural industry in the United States. Because flowers and pot plants are largely produced in plant-growing structures in temperate climates, floriculture is largely thought of as a greenhouse industry; there is, however, considerable outdoor culture of many flowers.

The industry is usually very specialized with respect to its crop; the grower must provide precise environmental control. Exact scheduling is imperative since most floral crops are seasonal in demand. Because the product is perishable, transportation to market must function smoothly to avoid losses.

The floriculture industry involves the grower, who mass-produces flowers for the wholesale market, and the retail florist, who markets to the public. The grower is often a family farm, but, as in all modern agriculture, the size of the growing unit is increasing. There is a movement away from urban areas, with their high taxes and labour costs, to locations with lower tax rates and a rural labour pool and also toward more favourable climatic regions (milder temperature and more sunlight). The development of airfreight has emphasized interregional and international competition. Flowers can be shipped long distances by air and arrive in fresh condition to compete with locally grown products.

The industry of landscape horticulture is divided into growing, maintenance, and design. Growing of plants for landscape is called the nursery business, although a nursery refers broadly to the growing and establishment of any young plant before permanent planting. The nursery industry involves production and distribution of woody and herbaceous plants and is often expanded to include ornamental bulb crops—corms, tubers, rhizomes, and swollen roots as well as true bulbs. Production of cuttings to be grown in greenhouses or for indoor use (foliage plants), as well as the production of bedding plants, is usually considered part of floriculture, but this distinction is fading. While most nursery crops are ornamental, the nursery business also includes fruit plants and certain perennial vegetables used in home gardens, for example, asparagus and rhubarb.

Next to ornamental trees and shrubs, the most important nursery crops are fruit plants, followed by bulb crops. The most important single plant grown for outdoor cultivation is the rose. The type of nursery plants grown depends on location; in general (in the Northern Hemisphere) the northern areas provide deciduous and coniferous evergreens, whereas the southern nurseries provide tender broad-leaved evergreens.

The nursery industry includes wholesale, retail, and mail-order operations. The typical wholesale nursery specializes in relatively few crops and supplies only retail nurseries or florists. The wholesale nursery deals largely in plant propagation, selling young seedlings and rooted cuttings, known as "lining out" stock, of woody material to the retail nursery. The retail nursery then cares for the plants until growth is complete. Many nurseries also execute the design of the planting in addition to furnishing the plants.

Bulb Crops

The bulb crops include plants such as the tulip, hyacinth, narcissus, iris, daylily, and dahlia. Included also are nonhardy bulbs used as potted plants indoors and summer outdoor plantings such as amaryllises, anemones, various tuberous begonias, caladiums, cannas, dahlias, freesias, gladioli, tigerflowers, and others. Hardy bulbs, those that will survive when left in the soil over winter, include various crocuses, snowdrops, lilies, daffodils, and tulips.

Many bulb crops are of ancient Old World origin, introduced into horticulture long ago and subjected to selection and crossing through the years to yield many modern cultivars. One of the most popular is the tulip. Tulips are widely grown in gardens as botanical species but are especially prized in select forms of the garden tulip (which arose from crosses between thousands of cultivars representing several species). Garden tulips are roughly grouped as early tulips, breeder's tulips, cottage tulips, Darwin tulips, lily-flowered tulips, triumph tulips, Mendel tulips, parrot tulips, and others. The garden tulips seem to have been developed first in Turkey but were spread throughout Europe and were adopted enthusiastically by the Dutch. The Netherlands has been the centre of tulip breeding ever since the 18th century, when interest in the tulip was so intense that single bulbs of a select type were sometimes valued at thousands of dollars. The collapse of the "tulipmania" left economic scars for decades. The Netherlands remains today the chief source of tulip bulbs planted in Europe and in North America. The Netherlands has also specialized in the production of related bulbs in the lily family and provides hyacinth, narcissus, crocus, and others. The Dutch finance extensive promotion of their bulbs to support their market. Years of meticulous growing are required to yield a commercial tulip bulb from seed. Thorough soil preparation, high fertility, constant weeding, and careful record keeping are part of the intensive production, which requires much hand labour. Bulbs sent to market meet specifications as to size and quality, which assure at least one year's bloom even if the bulb is supplied nothing more than warmth and moisture. The inflorescence (flowering) is already initiated and the necessary food stored in the bulb. Under less favourable maintenance than prevails in the Netherlands, a subsequent year's bloom may be smaller and less reliable; it is not surprising therefore that tulip-bulb merchants suggest discarding bulbs after one year and replanting with new bulbs to achieve maximum yield.

Herbaceous Perennials

Garden perennials include a number of herbaceous species grown for their flowers or occasionally used as vegetative ground covers. Under favourable growing conditions the plants persist and increase year after year. The biggest drawback to perennials as compared with annuals is that they must be maintained throughout the growing season but have only a limited flowering period. Typical perennials are hollyhocks, columbines, bellflowers, chrysanthemums, delphiniums, pinks, coralbells, phlox, poppies, primroses, and speedwells.

Perennials are often produced and sold as a sideline to other nursery activities; some are sold through seed houses. Perennial production could be undertaken on a massive scale, with attendant economies, but the market is neither large enough nor predictable enough (except for the greenhouse growing of such cut flowers as chrysanthemums and carnations) to interest most growers.

Shrubs

Production of ornamental shrubs is the backbone of the nursery trade in Europe and the United States. The nursery business is about equally divided between the production of (1) coniferous evergreens such as yew, juniper, spruce, and pine; (2) broad-leaved evergreens such as rhododendron, camellia, holly, and boxwood; (3) deciduous plants such as forsythia, viburnum, berberis, privet, lilac, and clematis; and (4) roses.

Fields of specialization have evolved within the ornamental shrub industry. Some firms confine activity mostly to production of "lining out" stock, which must be tended several years before reaching salable size.

The field grower may, in turn, specialize in mass growing for the wholesale trade only. The field plantings are tended until they attain marketable size. Because of the time required to produce a marketable crop and because of rising labour costs, this phase of the nursery industry involves economic hazards. But wholesale growing escapes the high overhead of retail marketing in urban areas, and, although many growers do sell stock at the nursery, they generally avoid the expensive merchandising required of the typical urban-area garden centre. Growers are especially interested in laboursaving technology and are turning to herbicidal control of weeds and shortcut methods for transplanting.

There is a well-established trade in container-grown stock—that is, nursery stock grown in the container in which it is sold. This practice avoids transplanting and allows year-round sales of plant material.

Roses

The production of roses is probably the most specialized of all shrub growing; the grower often deals solely in rose plants. Most are bud-grafted onto rootstocks (typically Rosa multiflora). This is the only way to achieve rapid and economical increase of a new selection to meet market demands. Large-scale production of roses has tended to centre in areas where long growing seasons make rapid production possible.

Because the budding operation calls for skilled hand labour and because field maintenance is expensive, few economies can be practiced in the production of roses. But distribution techniques that do offer certain economies have been developed. These include covering the roses with coated paper or plastic bags instead of damp moss to retain humidity and applying a wax coating to stems of dormant stock to inhibit desiccation.

Trees

Ornamental shade trees are usually grown and marketed in conjunction with shrubs. The 20th-century migration of people in many countries to suburban areas, coupled with the construction of houses on cleared land, has made shade trees an increasingly important part of the nursery trade. As interest in shade and ornamental trees increased, creation of improved cultivars followed. There is still some activity in transplanting native trees from the woodlot, and some are still grown from genetically unselected seed or cuttings; but more and more, like roses and shrubs before them, trees are vegetatively propagated as named cultivars, and many are patented.

The design and planning of landscapes has become a distinct profession that in many cases is only incidentally horticultural. Landscape architecture in its broadest sense is concerned with all aspects of land use. As a horticulturist, the landscape architect uses plants along with other landscape materials—stone, mortar, wood—as elements of landscape design. Unlike the materials of the painter or sculptor, plants are not static but change seasonally and with time. The colour, form, texture, and line of plants are used as design elements in the landscape. Plant materials are also manipulated as functional materials to control erosion, as surface materials, and for enclosures to provide protection from sunlight and wind.

Landscape architecture originated in the design of great estates, and home landscape is still an integral part of landscape architecture. More recently, however, landscape architecture has begun to include larger developments such as urban and town planning, parks both formal and "wild," public buildings, industrial landscaping, and highway and roadside development.

Floriculture

Floriculture refers to farming, plant care, propagation, and cultivation with one goal in mind, the maximum production of flower buds and flowers. Growers who focus on floriculture also generally experiment with creating new strains, cultivars, and varieties to improve bud and flower development.

Floriculture is an entire gardening spectrum that is geared towards understanding and improving all aspects of bud and flower creation, including indoor lighting, growroom requirements, greenhouse needs, plant nutrition, irrigation, pest management, and breeding new cultivars/strains. The goal of floriculture is always to improve the plant so it yields larger buds, more abundant buds, and optimal flowering times.

Growing with a floriculture objective means having a strong focus on the plant's spacing, pruning, ideal flower harvest time frame, and post-harvest chores such as storage and packaging of buds, flower heads, and other parts of the plant.

Floriculture encompasses all realms of successful growing, growth habits, and harvesting of a flowering plant. Growers usually center their goals around the plant's health, branching, growth size, bud formation, flowering, harvest, the plant's distinct desirable characteristics, and its overall flower and bud yield at the time of harvest.

All plants have two stages-the vegetative stage and the flowering stage. Floriculture singles out the flowering stage of the plant as being the most important aspect of the plant's life.

Floriculture growers work to make the plant's transition from the vegetative to the flowering stage an easy change in the hopes of boosting the plant's bud and flower growth to greater and newer heights.

Landscape Gardening

Landscape gardening is an aesthetic branch of Horticulture which deals with planting of ornamental plants in such a way that it creates a picturesque effect. It is a very fascinating and interesting subject. There are several definitions and expressions to define this subject. According to

Chambers' dictionary, the definition of landscape is the appearance of that portion of land which the eye can view at once and landscape gardening is the art of laying grounds so as to produce the effect of a picturesque landscape. Landscape gardening can be defined as the decoration of a tract of land with plants and other garden materials so as to produce a picturesque and naturalistic effect in a limited space. So landscape may or may not include plants. According to Bailey, Landscape gardening is the application of garden forms, methods and materials to the improvements of the landscape and the landscape in this connection is any area large or small on which it is desirable to develop a view or design.

Landscape gardening can also be defined as the beautification of a tract of land having a house or other object of interest on it. It is done with a view to create a natural scene by the planting of lawn, trees and shrubs. Landscape gardening is both an art and science of the establishment of a ground in such a way that it gives an effect of a natural landscape. It can be also defined as the imitation of nature in the garden. It can also be defined as improving of total living environment for the people. The expression of landscape may be gay, bold, retired, quiet, etc. This expression will conform to the place and the purpose. It should be a picture and not a collection of interesting objects.

Since the landscape gardening is the making pictures on the ground with plant and other material, landscape designer should be proficient in art, ornamental gardening, ecology and physiology. He should be an architect and engineer to appreciate the relationship between plant form, colours and buildings.

Natural Elements of Landscape

Different types of landscape depending upon prevailing geographical and agroclimatic conditions characterize Earth. There are mountains, hills, glens, valleys, seas, rivers, forests, plains, deserts, lakes, swamps, streams, etc. which comprise major part of natural landscape. At certain points, there is harmony between natural elements like ground forms, vegetation and even animal life. The landscape of such place is beautiful and conveys the feeling or mood of the landscape character like exhilaration, sadness, ceriness or awe. There are many qualities of natural landscape beauty like the picturesque; the ethereal, the serene, the delicate, the idyllic, the graceful, the majestic, the bold etc.

Man has copied the natural elements for improving landscape around him and converted certain areas in the form of garden for his pleasure.

Landscaping Principles

General principles of landscaping are as follows:

1. The ideal landscape garden is like ideal landscape painting which expresses some single thought or feelings. Its expression may be gay, bold, retired, quiet, etc.

2. Beauty and utility should be harmoniously combined.

3. Area should be divided into different parts and plan should be conceived for each area. Overall plan should be such that the observer catches the entire effect and purpose of the plan without stopping to analyse its parts.

4. Simplicity of design should be aimed at in the execution of the plan.

5. An ideal landscape should have open space.

6. Let the garden and building merge into each other. There should not be stopping abruptly particularly in front of building. The view of garden from the windows and doors should be very attractive. Planting around the building, climbers against wall and on the porch, decoration of verandah and rooms with attractive foliage, flowering plants, hanging baskets serve to unique the building with garden. Every part of the compound should be planned in such a way that it gives surprising effect to visitor.

7. Over crowding of plants and objects should be avoided.

Factors affecting the Landscape Design

There are several factors which affect the making of suitable design for particular site. These factors are:

1. Human choice: Man's ultimate desire is to make his living pleasurable and surroundings confortable. His dominance in making designs and selection of plant material is very well evident. Therefore, different styles of gardening have come into existence.

2. Site: This is an important factor and according to site, suitable design is made. In formal style gardening, the site is selected according to plan. Topography of the site also affects the design.

3. Views: Distant views of mountains, hills, woods, valley, etc. are preferred from the place of garden.

4. Heritage: One inherits the knowledge of botany and aesthetic sense and uses accordingly. Our rich heritage teaches us to use flowers and fragrant trees to improve the surroundings.

5. Climate: The climate of particular place affects the selection of plant material accordingly. Ideally suited plant material according to climate should be selected.

6. Soil: According to characteristics of soil types, suitable plants should be selected.

Garden Styles

Garden styles have been changed from time to time with the new ideas and necessities. Broadly, the styles of gardening are grouped into three categories i.e. (i) Formal style, (ii) Informal style and (iii) Free style of gardening

1. Formal style: Main features of this style of gardening are- First plan is made on the paper and then land is selected accordingly. Plan is symmetrical. These types of gardens are of geometric design i.e. squarish or rectangular. Therefore, the roads are cut at right angle. It has some sort of enclosure. Flower beds are also of geometric shapes. The arrangements of tree and shrubs are necessarily geometrical and kept in shape by trimming and training. Other features like fountains, water pools, cascades, etc. are used for further attraction. The examples of such style of gardening are Persian gardens and Mogul gardens.

2. Informal Style: This style reflects naturalistic effect of total view and represents natural beauty. This style is just contrast of above formal style. In this, plan is asymmetrical and according to the land available for making garden. Roads, paths are made curvaceous and bending. Water bodies are made of irregular shapes. Hillocks are made to create natural mountainous scenery. Flower beds are made of irregular shapes suiting to surroundings. Plants are allowed to grow in natural form and instead of trimming, annual pruning is done. Japanese gardens are the best example of this style of gardening.

3. Free style of gardening: This style combines the good points of both formal and informal style of gardening. Rose garden of Ludhiana is an example of this style of gardening.

Art Principles of Landscape

Landscaping is making of pictures with plant material and, hence, its principles are same as those of art. They are as follows:

*   Rhythm: Repetition of same object at equidistance is called rhythm. It can be created through the shapes, progression of sizes or a continuous line movement, rhythm creates movement to the eye. In gardens, generally trees of single species of equal height and shape are planted to create this effect. In Mogul gardens, fountains and water canals have also been extensively used to create such effect. Now-a-days other objects like lights are also used to create the effect of rhythm.

*   Balance: It is very important to maintain the balance on both sides of the central line. The principle involved in making balance of see-saw game can help in understanding this. Equal weights can be balanced only when they are equidistant from the centre. If weights are unequal, the heavier must move towards centre for making balance. The balance may be formal, informal or symmetrical types. Imbalance will look lopsided and will distract the attention. In making the balance with the plants, their form, colour, texture etc. are kept in view.

*   Accent or emphasis: The accent or emphasis is created in the gardens to avoid the monotonous view. It is the method to stress the most important thing. This also serves as the centre of attraction. Mostly unusual objects like tall fountain, tree, statue etc. are used to create the effect of accent or emphasis. In English gardens, statues have been used extensively to create such effects.

*   Contrast: This principle is most useful in emphasising the best features of an object. It can be very easily understood by following contrast colour theory. Against green background, a fleck of scarlet colour will make a contrast and will make scarlet colour prominent. In nature this is very common. Other contrast colour can also be used. Similarly, weeping growth habit against upright growth, dwarf against tall, rough texture against soft texture etc, are some of the examples which can be followed. It is also very important that one of the two contrasting objects must clearly dominate each other. In this way, one becomes feature whereas other acts as supporting background. The contrasting elements of equal power may create visual tensions.

*   Proportion: It is the relation of one thing to another in magnitude. When two or more objects are put together the proportions are established. In a landscape design, space provided

for lawn, paths, herbaceous borders, shrubbery border, trees, buildings and other garden objects should be in a right proportion. It will create harmonious effect and look better. Such effects can be noticed in Persian and Mogul gardens. Out of proportion allotment of area in garden will distract the attention. Proportion helps in space organization.

- Harmony: It is an overall effect of various features, styles, and colour schemes of the total scene. The degree of harmony or unity of various elements of landscape is a measure induced in us and is called as beauty. Therefore, the beauty can be defined as the evident relationship of all parts of a thing observed. When different parts of landscape are correctly placed in right way, produces a harmonious effect. Such landscapes create picturesque effect and appeals to visitors. On the contrary, the absence of harmony or lack of unity is ugliness.

Some Important Landscape Gardening Terms

- Axis: It is an imaginary line, which divides garden into two parts. This also connects two or more points. It is presented in the form of a path, line of fountains or trees, etc. If this axis divides garden into two equal parts, it is called central axis. In formal style, axis is central whereas in informal style, it is oblique. This axis controls the movement in the garden from the entrance to the terminal. An axis in garden is directional, orderly or dominating.

- Symmetrical plan: In this plan, the different objects are in equilibrium about a central point or on either side of an axis. Symmetrical plan or formal plan is synonymous of beauty and is pleasant and handsome. This is because the symmetry is to be associated with plan clarity, rhythm, balance, unity, etc. Symmetrical plan being precise and disciplined, it requires precision in detail and maintenance and bold in concept.

- Dynamic Symmetry: In such symmetrical plan, each pole generates its own magnetic field and between these two fields there is a field of dynamic tension.

- Asymmetrical Plan: In such plans, there is absence of symmetry on both the sides of axis but balance, unity and harmony are maintained.

- Circulation in Landscaping: In landscape gardening it means a pathway from entrance to terminal point. Circulation varies with the style of gardening and topography. The more of circulation patterns, it has more points of views and attraction. Common circulation patterns are depicted.

- Vista: It is a three dimensional confined view of terminal building or dominant element of feature. It may be natural or man-made. Natural vistas are very common around the lofty mountains and snowy peaks. Overall effect of vista may be of its characteristics. It may be calm or induce motion. There are three different parts of a vista viz:

    ○ It should be subject to a close control.

    ○ It should have a viewing station to see object or objects.

    ○ It should have intermediate ground. These three should be satisfactorily united and thus result into an effect of totality.

Olericulture

The area of horticulture that involves the production of vegetable food crops is olericulture. Olericulture includes the planting, harvesting, storing, processing, and marketing of vegetable crops. Sweet corn, tomatoes, snap beans, and lettuce are examples of vegetable crops.

Arboriculture

Arboriculture is the cultivation of trees, shrubs, and woody plants for shading and decorating. Arboriculture includes propagating, transplanting, pruning, applying fertilizer, spraying to control insects and diseases, cabling and bracing, treating cavities, identifying plants, diagnosing and treating tree damage and ailments, arranging plantings for their ornamental values, and removing trees. The well-being of individual plants is the major concern of arboriculture, in contrast to such related fields as silviculture and agriculture, in which the major concern is the welfare of a large group of plants as a whole.

The basic principles and objectives of arboriculture are of ancient origin. Early Egyptians transplanted trees with a ball of earth and originated the practice of shaping the soil around a newly planted tree to form a saucer to retain water, both still practiced. About 300 BC the Greek philosopher Theophrastus wrote Peri phytōn historia ("Inquiry into Plants"), in which he discussed transplanting of trees and the treatment of tree wounds. Virgil's Georgics portrays Roman knowledge of tree culture.

Trees or plants may be propagated by seeding, grafting, layering, or cutting. In seeding, seeds are usually planted in either a commercial or home nursery in which intensive care can be given for several years until the plants are of a size suitable for transplanting on the desired site. In soil layering, the shoots, or lower branches of the parent plant, are bent to the ground and covered with moist soil of good quality. When roots have developed, which may require a year or more, the branch is severed from the parent and transplanted. In an alternative technique, air layering, the branch is deeply slit and the wound covered by a ball of earth, moss, or similar material. The ball, enclosed in a divided pot supported from underneath, or in a sturdy paper cone, is kept moist. As in soil layering, the branch is severed and transplanted after roots have developed. Root cuttings can be used for propagating trees that do not normally produce roots from stems. Tree species such as willow and poplar that sucker, or send up shoots readily, are usually propagated from stem cuttings. Cuttings are made from deciduous plants during dormancy, preferably from the terminal growing shoots of the current season. Pieces 6 to 10 inches (15 to 25 centimetres) long with two or more buds are tied in bundles and stored in damp sand or moss for callus formation before planting in prepared beds. Root formation may be stimulated by application of growth-promoting chemicals or growth hormones.

In treating tree-trunk wounds in which large areas of bark are torn away, the bark around the wound is trimmed back to sound tissue and, at the top and bottom of the injury, trimmed to form a pointed ellipse of the wound area. The exposed wood is covered with wound dressing material, protecting it from wood decay fungi.

Flexible cables (guys) or rigid braces are used to support recently transplanted trees until the roots become established, or to lessen the danger that a tree with a weakened root system will be blown over by the wind; bracing is also used to support unduly long or heavy branches, to prevent splits developing at branch forks, or to permit healing of splits already developed.

Cavities in trunks, caused by decay-inducing fungi, may be treated with antiseptic dressing and left open, with drains installed at the bottom, or filled with concrete or other material after removal of the decayed wood.

Tree Shaping

Tree shaping (also known as pooktre, arborsculpture, tree training, and by several other alternative names) is the practice of training living trees and other woody plants into artistic shapes and useful structures. There are a few different methods of achieving a shaped tree, which share a common heritage with other artistic horticultural and agricultural practices, such as pleaching, bonsai, espalier, and topiary, and employing some similar techniques. Most artists use grafting to deliberately induce the inosculation of living trunks, branches, and roots, into artistic designs or functional structures.

Tree shaping has been practiced for at least several hundred years, as demonstrated by the living root bridges built and maintained by the Khasi people people of India. Early 20th century practitioners and artisans included banker John Krubsack, Axel Erlandson with his famous circus trees, and landscape engineer Arthur Wiechula. Contemporary designers include "Pooktre" artists Peter Cook and Becky Northey, "arborsculpture" artist Richard Reames, and furniture designer Dr. Christopher Cattle, who grows "grownup furniture".

Methods

There are various methods to achieving a shaped tree. These process use a variety of horticultural, arboricultural, and artistic techniques to craft an intended design. Chairs, tables, living spaces and art may be crafted from growing trees. Some techniques used for shaping trees are unique to a particular process, whereas other techniques are common to all, though the implementation may be for different reasons. These methods all start with an idea of the intended outcome. Some practitioners start with detailed drawings, or designs, other artists start with what the tree already has. Each process has it own time frame and a different level of involvement from the tree trainer. Some of these processes are still experimental, whereas others are still in the research stage. The trees might then either remain growing, as with the living Pooktre garden chair, or perhaps be harvested as a finished work like John Krubsack's chair.

Aeroponic Culture

The oldest known living examples of woody plant shaping are the aeroponically cultured living root bridges built by the ancient War-Khasi people of the Cherrapunjee region in India. These are being maintained and further developed today by the people of that region. Aeroponicgrowing was first formally studied by W. Carter in 1942, before the process had an English language name. Carter researched air culture growing and described "a method of growing plants in water vapor to facilitate examination of roots". Later researchers, including L. J Klotz and G. G. Trowel, expanded on his work. In 1957, F. W. Went described "the process of growing plants with air-suspended roots and applying a nutrient mist to the root section," and in it he coined the word 'aeroponics' to describe that process. In 2008, root researcher and craftsman Ezekiel Golan described and secured a patent for a process which allows the roots of some aeroponically grown woody plants to lengthen and thicken while still remaining flexible. At lengths of perhaps 18 ft or more, the soft roots can be formed into pre-determined shapes which will continue thickening after the shapes are formed and as they continue to grow. Newer techniques and applications, such as eco-architecture, may allow architects to design, grow, and form large permanent structures, such as homes, by shaping aeroponically grown plants and their roots.

Treenovation created this chair using the techniques of Aeroponic root shaping.

Instant Tree Shaping

Arborsculpture bench by Richard Reames created using the techniques as described in his books How to grow a chair and Arborsculpture.

Instant tree shaping starts with more mature trees, perhaps 6–12 ft. long and 3-4in in trunk diameter, which are woven into the desired design and held until cast. Understanding a tree's fluid dynamics is important to achieving the desired result.

Bending is sometimes used to achieve a design. If a plant's tissue is bent at too sharp an angle it may break, which can be mostly avoided by un-localizing the bend. This is achieved by making small bends along the curve of the tree. Bends are then held in place for several years until their form is permanently cast. The tree's rate of growth determines the time necessary to overcome its resistance to the initial bending. The work of bending and securing in this way might be accomplished in an hour or perhaps in an afternoon depending on the design.

Ring barking is sometimes employed to help balance a design by slowing the growth of too-vigorous branches or stopping the growth of inopportunely placed branches, using different degrees of ring barking, from simple scoring to complete removal of a 3/8 in-wide band of bark.

Creasing is folding trees such as willow and poplar over upon themselves, creating a right angle. This method is more radical than bending.

With this method it is possible to perform initial bending and grafting on a project in an hour, as with *Peace in Cherry* by Richard Reames, removing supports in as little as a year and following up with minimal pruning thereafter.

Gradual Tree Shaping

"Grownup furniture" by Chris Cattle created using a gradual tree shaping method.

Gradual tree shaping starts with designing and framing. These are fundamental to the success of the piece. Once these are set up, young seedlings or saplings 3–12 in. long are planted.

The training starts with young seedlings, saplings or the stems of trees when they are very young, and are gradually shaped while the tree is growing to form the desired shape. There is a small area just behind the growing tip that forms the final shape. The shaping zone, it is the shaping of this area requires day to day or weekly guiding of the new growth. The growth is guided along predetermined design pathways; this may be a wooden jig or complex wire design.

With this method the time frame is longer than the other methods. A chair design might take 8 to 10 years to reach maturity Some of Axel Erlandson trees's took as long as 40 years to assume their finished shapes.

Common Techniques

Grafting

Grafting is a common technique used by all the different methods. Grafting exploits the natural biological process of inosculation. Grafting is where a branch or plant is cut and a piece of another plant is added and held in place. There are various types of grafting, in all types the idea is to encourage the tissues of one plant to fuse with those of another.

Grafting is applied to create permanent connections and joins. In some cases the trees are grafted while they are growing in others the mature trees may be intertwining and then grafting together the stems of two or more trees in order to create chairs, ladders, and other fanciful sculptures.

Framing

Framing may be used for various purposes and might consist of any one or a combination of several materials, such as timber, steel, tubes made of hollow out trees, complex wire designs, or the tree itself living. It can be used in many project designs to support grafted joints until the grafts are well-established. Some process might employ framing to hold a shape created by bending or fletching mature trees until the tissues have overcome their resistance to the initial bending and grown enough annual rings to cast the design permanently. Others might use framing to support and shape the growth of young saplings until they are strong enough to maintain an intended shape without support.Still other approaches might employ frames to guide the roots of aeroponically grown trees into desired shapes.

Pruning

Pruning can be used to balance a design by controlling and directing growth into a desired shape. Pruning above a nodecan steer plant growth in the direction of the natural placement of that leaf bud. Pruning may also be used to keep a design free of unwanted branches and to reduce canopy size. Pruning is sometimes the only technique used to craft a project. Deciduous trees are mainly pruned in winter, while they are dormant above-ground, although sometimes it is necessary to prune them during the growing season. Trees repeatedly subjected to hard pruning may experience stunted growth, and some trees may not survive this treatment.

Pleaching

Pleaching is a technique used in the very old horticultural practice of hedge laying. Pleaching consists of first plashing living branches and twigs and then weaving them together to promote their inosculation. It is most commonly used to train trees into raised hedges, though other shapes are easily developed. Useful implementations include fences, lattices, roofs, and walls. Some of the outcomes of pleaching can be considered an early form of what is known today as tree shaping. In an early, labor-intensive, practical use of pleaching in medieval Europe, trees were installed in the ground in parallel hedgerow lines or quincunx patterns, then shaped by trimming to form a flat-plane grid above ground level. When the trees' branches in this grid met those of neighboring trees, they were grafted together. Once the network of joints were of substantial size, planks were laid across the grid, upon which they built huts to live in, thus keeping the human settlement safe in times of annual flooding. Wooden dancing platforms were also built and the living tree branch grid bore the weight of the platform and dancers.

In late medieval European gardens through the 18th century, pleached allées, interwoven canopies of tree-lined garden avenues, were common.

Structure

Living grown structures have a number of structural mechanical advantages over those construct-ed of lumber and are more resistant to Leaf decomposition. While there are some decay organisms that can rot live wood from the outside, and though living trees can carry decayed and decay-ing heartwood inside them; in general, living trees decay from the inside out and dead wood decays from the outside. Living wood tissue, particularly sapwood, wields a very potent defense against decay from either direction, known as CODIT. This protection applies to living trees only and var-ies among species.

Growing structures is not as easy as it would seem. Quick growing willows have been used to grow building structures, they provide support or protection. A young group of German architects are in the process of such a structure and they are continually monitored and checked. Once the trees are of age to be able to take on load-bearing weight they are tested for stability and strength by a structural engineer. Once this is approved the supporting framework is removed. Projects are lim-ited to the trees' weight loading ability and growth. This is being studied and the load capacity will be proved by testing on prototypes.

Design Options

Becky's Mirror by Pooktre

Designs may include abstract, symbolic, or functional elements. Some shapes crafted and grown are purely artistic; perhaps cubes, circles, or letters of an alphabet, while other designs might yield any of a wide variety of useful shapes, such as clothes hangers, laundry and wastepaper bins, ladders, furniture, tools, and tool handles. Eye-catching structures such as living fences and jun-gle gyms can also be grown, and even large architectural designs such as live archways, domes, gazebos, tunnels, and theoretically entire homes are possible with careful planning, planting, and culturing over time. The Human Ecology Design team (H.E.D.) at the Massachusetts Institute of Technology is designing homes that can be grown from native trees in a variety of climates.

Suitable trees are installed according to design specifications and then cultured over time into in-tended structures. Some designs may use only living, growing wood to form the structures, while others might also incorporate inclusions such as glass, mirror, steel and stone, any of which might be used either as either structural or aesthetic elements. Inclusions can be positioned in a project as it is grown and, depending on the design, may either be removed when no longer needed for support or left in place to become fixed inclusions in the growing tissue.

Pomology

Pomology is the planting, harvesting, storing, processing, and marketing of fruit and nut crops. Fruit crops include both large and small fruits. Examples of large fruits are peaches, apples, and pears. Small fruits include strawberries, raspberries, and blueberries. Almonds, pecans, and walnuts are popular nut crops.

Viticulture

Viticulture is the process of grape production. Grapes are grown around the globe apart from the Antarctica, and they have high adaptability properties to different environments. Grapes are fruits which are used to produce wine. The people who study the science behind grape production are called viticulturists. Viticulturists study pests and diseases to control, irrigation, fertilization, fruit development, and characteristics of grapes. They are also responsible in managing the canopy on which the grapes grow, when to prune, and when to harvest the fruit. Winemakers also liaise with viticulturists in wine production because the best vines give the best wine. Different varieties of grapes are now approved by the European Union as the real grapes for wine production, because of their characteristics. Some of the wines produced using grapes include the red wine, which is produced from gulp of black and red grapes. Grapes are also used in making raisins.

Uses of Grapes

Grapes are berries- The system through which grapes are organized is called clustering. Some clusters contain grapes which are compact berries while others are spread out. Long clustered grapes spread out while short clusters are packed together. Some grapes ripen together making harvesting easier while others ripen individually within the same cluster. Raisins are made from dried grapes which could be seedless and they are used in cooking, brewing, and baking. Fertilization occurs in one to four seeds within a berry. Inside each grape, there is a rachis which allows the grape to get nutrients and water. One plant can produce 100 to 200 grapes. The skin of grape weighs about 5% to 20% of the total weight of the grape. The skin when ripe contains tannin and some aromatic substances. When the grape is ripe, the tannin is used to formulate color and body shape.

Conditions for Growing Grapes

Each type of grape has a favorable climatic condition. Temperatures and rainfall are uncontrollable conditions and each year will see the growth of unique grapes. During summer times, the right temperatures help in the ripening of the grapes. These cool temperatures allow for the resting of the fruit. Very cold temperatures cause frosting of the fruit. The grapevine should also be on a hillside, and this allows for sunrays falling at an angle where all vines get sunshine, unlike on flat ground where the sunrays do not reach every plant well. Slopes also offer better drainage. Quality soil is important in determining the health and growth of the vine. The right kind of soil allows for the plant's roots to develop well. Soil with loose texture, moderate fertility, and great drainage are favorable for grape growth.

Problems facing Grape Farming

Weather conditions like wind, frost, a lot of heat, and poor drainage are some of the factors that adversely affect the quality grape harvest. Diseases are also rife among the grape farmrs. Some of the viticulture hazards include oidium, downy mildew, fan leaf, plant virus, and phylloxera. Spraying the plants with copper sulfate treats the problem of downy mildew. Some species from North America have developed resistance to this disease. Fan leaf has no cure so affected are removed. Oidium can be terminal and does well in the shade and cooler temperatures.

Grape Harvesting

A green harvest is done on immature grapes to reduce yield. Removing those green grapes helps the remaining ones to put their energy into developing better and healthy grapes. A healthy, vigorous and more mature flavor compound is achieved through removing some bunches of green grapes. The practice of intermixing different varieties of grapes produces a wine called a field blend. Field blend allows for effortlessly blending of different varieties with different genetic makeup. In Germany, field blend is called a mixed set.

Oenology

Oenology is the science of all aspects of winemaking and wine. Oenology does not study vine growing or grape harvesting, those areas fall under the subfield of viticulture. A person who studies oenology is called an oenologist. The term oenology comes from the Greek word oinos, which means "wine" and logio, which means "study of".

Oenology focuses on the study of the wine making process from crushing to serving and everything in between, including the must, fermentation, racking, bottling, and aging of wine. Specifically, oenology studies the science of the complicated chemical reactions that take place in the grape juice as it becomes wine.

Plant Propagation

Plant propagation can also refer to the artificial or natural dispersal of plants. Plant propagation is the process of creating new plants from a variety of sources: seeds, cuttings, bulbs and other plant parts. Basically all plants in this universe multiply themselves both by sexual (seeds) or asexual (vegetative) means. However man has now developed some more techniques for speedy and better multiplication of these plants.

Plant Propagation Methods

The first and foremost objective of a propagation method should be to produce individuals that are identical to mother or original plant. Thus a successful propagation method is one which transmits all desirable characters of a mother plants to the offspring's.

1. Sexual Propagation: The plants which multiply through seeds as a mode of perpetuation. Example: Most annuals, biennials and many perennial fruit plants; vegetable crops; plantation, aromatic and medicinal plants; ornamental flowering and shade providing shrubs and trees etc. In general the self pollinated plants which are considered homozygous are propagated by seeds. In sexual method of propagation, the sex organs of flower are involved in process like pollination and

fertilization leads to the formation of seeds. Seeds are typically produced from sexual reproduction within a species may have different characteristics from its parents.

Example: Papaya, phalsa, mangosteen, rootstocks plants of many fruit crops raised through sexual method of propagation.

2. Asexual/Vegetative Propagation Propagation by apomictic seedlings: Apomictic seedlings are identical to their mother plants, and similar through the plants raised through other vegetative means, as it has the same genetic makeup as that of the mother plants. Example: Citrus, mango, apple etc.

3. Propagation by vegetative structure: It is also called as vegetative propagation as it involves only vegetative parts without any sexual plant parts. The plant parts like leaf, stem, root and other root producing plant organs are used. The new individual propagated through this method is true to type. The commercially important fruit crops are propagated by vegetative method.

Types of Plant Propagation

Propagation by Cuttings

A cutting is a piece of vegetative tissue (stem, root or leaf) that, when placed under suitable environmental conditions, will regenerate the missing parts and produce a self-sustaining plant. In this method of propagating fruit plants, a plant (generally stem) having at least few buds, when detached from parent plant and placed under favourable conditions develop into a complete plant resembling in all characteristics to the parent from which it was taken. This method is commonly used in plants, which root easily and readily, thus multiplication of plants is very quick and cheap. The fruit plants like Phalsa, pomegranate, lemon and grapes etc. are commercially propagated by cuttings. However this method of propagation has certain disadvantages like:

1. Some plants do not root easily.

2. The desired benefit of rootstock of rootstock on a desired variety cannot be exploited. Cuttings are prepared from vegetative portion of the plants such as stem, root, leaf and are classified according to the plant part used and are described below:

Stem Cutting

A stem cutting is any cutting taken from the main shoot of a plant or any side shoot growing from the same plant or stem. Propagation by stem cuttings is the most commonly used method to propagate many woody plants. There are several types of stem cuttings:

1. Hardwood cutting: Hardwood cuttings are prepared during dormant season, usually from one year old immature shoots of previous season's growth. In case of deciduous plants, the cuttings are made after pruning. Generally the cuttings of 15-20 cm length and having 3-5 buds are preferred depending upon species. While preparing the cutting, a straight cut is given at the base of shoot below the node while a slanting cut 1-2 cm above the bud is given at the top, e.g. grape, fig. pomegranate, mulberry, kiwifruit, olive, quince, hazel nut, chest nut, plum, gooseberry and apple.

2. Semi Hardwood Cutting: Semi hardwood cuttings are prepared from partial matured, slightly

woody shoot. These are succulent and tender in nature and are usually prepared from growing wood of current season's growth. The length of cutting varies from 7-20 cm. The cuttings are prepared by trimming the cutting with straight cut below a node and removing a few lower leaves. However, it is better to retain two to four leaves on the top of cuttings. Treating the cutting with 5000 ppm IBA (an rooting hormone) before planting gives better result. The best time of taking cutting is summer when new shoots is emerged and their wood is partially matured. Ex. Mango, Guava, Jackfruit, Lemon etc.

3. Soft-Wood Cutting: Softwood cutting is the name given to any cuttings prepared from soft, succulent and non-lignified shoots which have not become hard or woody. Softwood cuttings are taken from new growth of the current season. They are still flexible and non-woody. Softwood cuttings are generally the easiest to root and don't require special handling. e.g. lime and lemon.

Leaf Bud Cutting

There are relatively few plants which can be reproduced by using a leaf or the portions of a leaf to produce a new plant. The reason for this is that leaf cuttings must regenerate both new root and bud tissues. Not many plants have this capacity. Most of the plants propagated by leafcuttings are house plants with thick, fleshy leaves. Depending on the plant to be propagated, the leaf cutting may involve the leaf blade only, or a leaf with petiole, or merely portions of a leaf. Leaf-bud cuttings are unlike leaf cuttings in that it contains a portion of stem tissue and most importantly, a bud. The bud, located at the junction of the leaf petiole and the stem, is a preformed growing point. Leaf cutting should preferably be prepared during growing season because buds if inter in dormancy may be difficult to force to active stage. Plant material selected for leaf cuttings should be healthy, actively growing and free of insect or disease problems. e.g. Blackberry, lemon, raspberry etc.

Root Cutting

Root cuttings can be used to propagate a range of herbaceous perennials in late autumn or early winter when the plants are dormant. Root cuttings are used to propagate plants that naturally produce suckers (new shoots) from their roots. Root cuttings require no special aftercare needed for aerial cuttings as well as large numbers of new plants can be generated from each parent plant and the plants derived from root cuttings are relatively large and vigorous. Another advantage of root cuttings is that plants from root cuttings are free of foliar pests and pathogens that might affect their parents, such as stem and leaf nematodes, e.g. blackberry, fig, cherry, raspberry etc.

Propagation by Layering

Air-layering

In this method one-year-old, healthy and straight shoot is selected and ring of bark measuring about 2.5 to 4.0 cm just below a swollen bud is removed. The cut is then surrounded sphagnum moss or any other material which can retain moisture for long period of time, can be used for this purpose and is wrapped with a polyethylene strip (200–400 gauge). Both ends are tied with fine rope or rubber bands to make it practically air-tight and after 30 to 45 days, roots are formed in the aerial part of the plant. The ideal time is February–March and July–August months for air- layering. Litchi, Lime and sweet lime can be propagated by air layering. Application of root promoting hormones at the time of layering helps to get profuse rooting within a short time. Root promoting substances may be applied as powder or in lanolin or as a solution.

Trench Layering

In this method, it is important to establish a permanent row of plants to be propagated. A branch is laid horizontally in a small trench to encourage the development of several new shoots from it. As these shoots grow, soil is filled around them and roots eventually develop. The mother plants are planted at the base of a trench at an angle of 45 0 in rows. The long and flexible stems of these plants are pegged down on the ground to form a continuous line of layered plants. The young shoots then arise from these plants are gradually mounded up to a depth of 15-20 cm, e.g., Apple rootstocks (M16 and M25), cherry, plum etc.

Tip Layering

In this method of propagation, bending of a plant stem to the ground and covering the tip with soil so that roots and new shoots may develop. In tip layering, the tip of shoots are bend to the ground and the rooting takes place near the tip of current season's shoot.

Serpentine/Compound Layering

Serpentine layering, also known as compound layering is similar to simple layering, except that multiple sections of the stem are buried, resulting in multiple plants from one stem. With this technique, a long, vine-like stem is ideal. It is modification of simple layering in which one year old branch is alternatively covered and exposed along its length. e.g. American grape.

Mound Layering/Stooling

In this method, plants are headed back to 15-20 cm above the ground level during dormant season. The new shoots come out within two months after heading back. These sprouts are then girdled near base and rooting hormone (IBA), made in lanolin paste is applied to the upper portion of cut with moist soil. The rooting of shoots is observed within 25–35 days. The stooling is used in the commercial production of clonal rootstocks of temperate fruit crops like apple, pear etc. and sometimes in guava also.

Propagation by Grafting

Grafting is another method of vegetative propagation, where two plant parts are joined together in such a manner that they unite and continue their growth as one plant i.e. the stock and scion are placed in close contact with each other and held together firmly, until they unite to form a composite plant. The different methods of grafting are tongue grafting, cleft grafting, approach grafting, side grafting and veneer grafting etc.

Tongue Grafting (Splice/Whip Grafting)

It is the simple and popular propagation method used in apples and widely used in pear. This method is commonly used when the stock and scion are of equal diameter. Each scion sticks should contain at least two to three sets of buds. Identical cuts are made at the top of the rootstock and bottom of the scion, so the two pieces fit together nicely. About one-year-old rootstock is headed back at a height of 23-25 cm from the soil and a diagonal cut is made at the distal end of the rootstock. A similar slanting cut is made on the proximal end of the scion. Try to make this cut with one stroke of the knife. The cut surface of both rootstock and scion are bound together and tied firmly. The scion having 2 to 3 buds is then tightly fitted with the rootstock taking care that the cambium layer of at least one side of the stock and scion unites together. This is then wrapped with polyethylene strip. Tongue grafting is done in March–April in high hills and dry temperate zone while February-March in lower elevation. Many fruit plant are propagated by whip grafting.

Cleft Grafting

This is also known as wedge grafting. Wedge grafting is a relatively easy method of propagation. In this technique, proper selection and preparation of scion sticks is very important for obtaining higher success of graft. This method is useful in the nursery where the rootstock is quite thicker than scion and tongue grafting cannot be employed successfully. Rootstock up to 8 cm girth is selected for this purpose. The rootstock is cleft grafted after decapitating the stock 45 cm above the ground level. The beheaded rootstock is split to about 5cm deep through the center of stem. After that a hard wooden wedge is inserted to keep open for the subsequent insertion of scion. The scion of 15-20 cm size is taken from a terminal shoot, which is more than three month old and then it is wedge securely (6-7 cm). The cleft of the scion then slipped into the split of the stock. In thicker rootstock more than one scion should be inserted. Care must be exercised to match the cambium layer of the stock and scion along with full length of each component. This technology has overcome seasonal barriers and planting materials could be raised throughout the year either in greenhouse or under open field conditions. Polyethylene cap facilitates early sprouting and ensures good success rate of grafts, e.g. avocado, apple, pear, plum, mango etc.

Side Grafting

In this method of grafting, a three sided rectangular cut about 4.0 x 1.25 cm is given on the rootstock at a height of about 15-18 cm and the bark of the demarcated portion is detached from the rootstock. A similar cut is also given on the base of the scion stick to expose cambium. The scion should be prepared well in advance before the actual grafting. The healthy scion shoots of previously mature flush are selected. The selected scion shoots should have plump terminal buds. After the selection of the scion shoots, remove the leaf blades, leaving petioles attached to the scion. In about 8 to 10 days, the attached petioles drop automatically and terminal buds become swollen. Now the scion stick should be detached from the mother tree and grafted on the stock. The prepared scion is inserted under the bark flap of the rootstock after given a slanting cut so that the exposed cambium of the two components is in close contact with each other. The bark flap of the rootstock is resorted in its position. The graft union is then tied firmly with polythene strip. After the successful completion of the grafting operation, a part of the top of rootstock is removed to encourage growth of the scion. When the scion has shown emergence of leaves, the root stock portion above the graft union should be removed. Side grafting can be carried out successfully from March to October but success during the May and October is rather low. This method of propagation is commonly used in mango.

Inarching or approach Grafting

The method of inarching or approach grafting is quite cumbersome and time consuming, but it is still the leading method for commercial propagation of many fruit plants. It is generally used for repairing or replacing damaged root system and hence also called as repair grafting. This method of grafting is termed approach grafting, as the rootstock is approached to the scion, while it is still attached to the mother plant. Selection of parent tree for taking the scion is an important factor for its success. The scion plant should be healthy, vigorous and high yielding. The stock is brought close to the scion. In this method the diameter of rootstock and scion should be approximately the same. A slice of bark along with a thin piece of wood about 6-8 cm long and about 1/3 inch in thickness at height is removed from matching portions of both the stock and the scion. They are then brought together making sure that their cambium layers make contact with each other. These grafts are then tied firmly with polythene strip or any other tying material. The stock and scion plants are watered regularly to hasten the union. The union is complete in about 2 to 3 months. After successful union, stock above and scion below the graft union are looped of gradually. Last week of July or the first week of August is the best period for approach grafting, e.g. mango, sapota, guava, litchi etc.

Veneer Grafting

This method of propagation possesses promise for mass scale commercial propagation. It is simple method of propagation. However, several factors affecting the success like age and diameter of scion, season of grafting and defoliation period of scion stick etc. About one year old rootstock is suitable for this method. However, if the stock attains suitable thickness (about 1 to 1.5 cm dia) earlier than a year, can be utilized for rootstock purpose. Better success is obtained with a scion stick of 3-6 months of age with lust green leaves are selected. The scion sticks are pre-defoliated 5-10 days prior the grafting leaving the petiole attached for making the auxiliary and apical buds active. Usually, the terminal and next to terminal shoots are most ideal. For conducting this grafting operation, a downward and inward 25–35 mm long cut is made in the stock at a height of about 15-20 cm. At the base of cut, a small shorter cut is given to intersect the first so asto remove the piece of wood and bark. The scion stick is given a long slanting cut on one side and a small short-cut on another side to match the cuts with the rootstock. Now the scion is inserted in the stock so that the cambium layers come in contact from the longer side. The graft union is then tied with polyethylene strip. After the scion remains green for more than 10 days, the rootstock should be clipped in stages. When scion growth begins, the shoot of rootstock is removed above the graft union. eg Mango.

Softwood Grafting

The name indicates grafting is done on newly emerged rootstock of 40-60 day-old. This is easy and simple method and can be practiced throughout the year raised with 60-90% success. This method is useful for in situ grafting while establishing new orchards with already established rootstocks in the field. In this method, remove the leaves of the selected seedling using sharp knife, retaining one or two pairs of bottom leaves. Give a transverse cut on the top soft portion of the seedling. Make a cleft of 3-4 cm deep in the middle of the decapitated stem of the seedling by giving a longitudinal cut. A little wood is removed from the inner side of the cleft at the top. The scion is prepares by mend the cut end of the scion into wedge shape of 3-4 cm in length by chopping off the bark and little wood from the two opposite sides, taking care to retain some bark on remaining two sides. Insert the wedge of the scion into the cleft of the seedling taking care that cambium layers of stock and scion come in perfect contact with each other. Secure the joint firmly with polythene strip (30 cm long, 2 cm broad and 150 gauge). Cover the scion with polythene cap (15 cm x 10 cm, 100 gauge) and tie it at the bottom to maintain humidity and to protect the apical bud from drying. The cap should not touch the terminal bud. July and August months with high humidity and moderate temperature are the best for the success of softwood grafting. This is very famous and successful propagation method in cashew nut, mango, sapota etc.

Propagation by Budding

Budding is a method of propagation in which only one bud of desired scion is inserted in the root-stock. The ideal condition of budding is when the bark starts slipping both on the stock and scion. It indicates that the cambium is active. Important method of budding are shield or t- budding, patch budding, chip budding, ring budding etc.

Shield or T-budding

'T' budding or shield budding is a special grafting technique in which the scion piece is reduced to a single bud. A small branch with several buds suitable for T budding on it is often called a bud stick. Successful T budding requires that the scion material have fully-formed, mature, dormant buds, and that the rootstock be in a condition of active growth such that the bark is slipping indicating the vascular cambium is actively growing, and the bark can be peeled easily from the stock piece with little damage. Bud sticks having plump, healthy buds are suitable as scions. Leaf blades are clipped from the budsticks, leaving the petiole intact. Budding knives should be kept very sharp, so that as little damage as possible is done to the bud.

The bud and a small sliver of the wood underneath it are cut from the bud stick using an upward slicing motion. The cut should begin about 1/2 to 3/4 inch below the bud, and should go deep enough into the wood so that when the cut is finished about 1/2 to 3/4 above the bud, the bark and a small sliver of wood are cut off. A perpendicular cut across the top of the upward cut will separate it from the bud stick. Buds must be cut from the bud stick just prior to grafting, otherwise they will dry out. The bark is carefully slipped from the stem of the rootstock exposing a "pocket" into which the bud shield can be placed. Care should be taken not to tear the flaps of bark in the process of spreading them. If the bark does not slip easily, this indicates that the stock is not in active growth and the process should be conducted later when active growth has resumed. The bark flaps are held tightly against the bud as they are wrapped with a budding rubber, grafting tape or other suitable closure. This closure must be removed in 2 to 3 weeks after the union has healed. If the material does not break down, it will girdle the rootstock, e.g. Citrus, plum, peach, cherry, ber, rose etc.

Patch budding

A rectangular patch of bark is removed completely from the stock and replace with a similar patch of bark containing a bud of desired variety. Normally done during the growing season when the bark separates readily from the wood along the cambial layer. It is successfully used in species having thick bark such as walnut, pecan nut.

Chip Budding

Chip budding is successful method of budding when the bark of the stock does not slip easily. A chip of bark and wood is removed from the smooth surface between the nodes of the stock. A chip of similar shape and size is then removed from the bud wood of desired cultivar. For which, a 2-3 cm long down ward cut is made through the bark and slightly in to the wood of the stock. Then a second cut of about 2.5 cm is made so that it bisects the first cut at an angle of 30-45 °C in this way the chip of wood is removed from the stock. The bud chip then slipped in the place of rootstock from where chip has been removed, e.g. mango, grape etc.

Ring Budding

In ring budding, a complete ring of bark is removed from the stock and it is completely girdled. A similar ring of bark containing a bud is removed from the bud stick and is inserted on to the root-stock. In this budding both scion and stock should be of same size. It is utilized in peach, plum, ber, mulberry etc.

Tissue Culture/Micropropagation

The principles of tissue culture can be successfully employed in horticultural crops. Quite a large number of ornamental plants are reported to respond to propagation by tissue culture method. Few such plants are gladiolus, carnation, lily, rose, gerbera, anthurium, magnolia, fern, cacti, etc. Propagation of ornamental plants by this method is gaining popularity. Among fruit crops, tissue cultured plants are commercially produced in banana, strawberry etc. The supply of tissue cultured plants of Grand Nain variety of banana has revolutionised the banana industry.

Advantages of Vegetative Propagation

- The horticultural crops which do not produce viable seeds are propagated by vegetative method.
- Most of the important fruit crops are cross pollinated and are highly heterozygous. When propagated through seeds, the progenies shows large variation, so vegetative propagation is remedy for these crops.
- The asexual propagation method gives true to type plants.
- The vegetative way propagated plants bear fruits early.
- In case of fruit crops where root stocks are used, the root stocks impart insect or disease resistance to the plant.
- Vegetative propagation helps to alter the size of the plant. i.e. dwarfing effect. This helps for spraying, intercropping & harvesting of crops easy and economical.
- By grafting method different variety of fruit crop can be grown & harvested.
- Inferior quality fruit plants can be converted into good quality plants.
- By means of bridge grafting a repairing of injured plants can be done.

Disadvantages of the Vegetative Propagation

- By vegetative propagation new variety cannot be developed.

- It is an expensive method of propagation and required specialized skill.

- The life span of vegetatively propagated plants is short as compared to sexually propagated plants.

- As all the plants are homozygous the whole plantation may get attacked by a particular pest or disease.

- Viral diseases could be transferred through vegetative parts.

References

- Horticulture, science: britannica.com, Retrieved 12 May, 2019

- Floriculture: maximumyield.com, Retrieved 20 February, 2019

- Landscape, horticulture: indiaagronet.com, Retrieved 23 July, 2019

- Horticulture: senecahs.org, Retrieved 3 April, 2019

- Arboriculture, science: britannica.com, Retrieved 27 January, 2019

- What-is-viticulture: worldatlas.com, Retrieved 9 March, 2019

- Oenology: winefrog.com, Retrieved 29 June, 2019

- Propagation-in-Fruit-Crops: biotecharticles.com, Retrieved 19 April, 2019

Chapter 2

Olericulture: The Practice of Vegetable Farming

Olericulture is the science and practice of growing vegetables for food. It also deals with the production, storage, processing and marketing of vegetables. There are many vegetable crops which fall under this category such as potato, tomato, carrot and onion. The diverse applications of olericulture in the current scenario have been thoroughly discussed in this chapter.

Olericulture is one of the branches of Horticulture that deals with the vegetables. The word olericulture is derived from the Latin word Oleris which means pot herb and the English word culture which means cultivation. Thus olericulture means cultivation of pot herbs. However, in the present days, it is bradly used to indicate the cultivation of vegetables. The term vegetable gardening is more popular to signify olericulture in the present context.

Vegetable: The term vegetable is applied to the edible herbaceous plant or plant parts thereof, which are consumed generally in the unripe stage after cooking.

Importance of Vegetables in Human Nutrition

The balanced diet contain adequate energy source, nutrients and vitamins, mineras, carbohydrates, fats, protein etc.

Vegetable are the reliable source for many dietary factors. As vegetable contain many of the dietary factors like vitamins, minerals and amino acids they are considered as protective supplementary food. They produce taste, increase appetite and produce fair amount of fibres. They maintain good health and protect against degenerative diseases. They can neutralise the acids produced during digestion of proteins and fats.

Nutrients which are present in vegetables vary from crop to crop. Peas and beans are enriched with proteins. Root crops like Tapioca, Sweet potato and potato are well known for carbohydrates, calcium K, Fe are the important minerals which are lacking in cereals and these are available in abundant quantities in the vegetables like peas, beans, spinach and bendi.

Amaranth, cabbage, beans contain large quantity of cellulose which aid in digestion. All the leaf and fruit vegetables possess the required quantities of vitamins.

S. No	Dietary factors	Source Vegetables
1	Calories	Sweet potato, tapioca, yam, colacasia corms, potato, Brussels Sprouts, onion and garlic, immature seeds of broad bean and peas Phaseolus lunatus (Lima bean), Pussia fada (Broad bean).
2	Proteins	Peas, double bean, winged bean (Psochocarpus tetragonolobus), Garlic, Brussels sprouts, cowpea, lema bean seeds, amaranthus Leaves, drumstick leaves and menthe.

3	Vitamin A (Beta carotene)	Carrot, spinach, turnip green, palak, mustard green, amaranth, coriander, colacasia leaves, sweet potato, pumpkin, tomato
4	Vitamin B complex	Peas, broad bean, lema bean, garlic, asparagus, colacasia and Tomato.
5	Vitamin C	Turnip green, green chillies, Brussels sprouts, mustard green, Amaranth, coriander, drumstick leaves, cauliflower, knoll khol Spinach, cabbage, bitter gourd and reddish leaves.
6	Calcium	Curry leaves, amaranth leaves, drumstick leaves, menthi, turnip, Mustard green, coriander and palak.
7	Iron	Drum stick leaves and fruits, amaranth, menthi, mint, coriander, Spinach, palak and mustard green. Spinach, lettuce, cabbage
8	Roughages	Amaranth and root vegetables.
9	Vegetable milk	Pea pods and cabbage leaves.

Types of Vegetable Gardens

Vegetable gardens can be classified into 7 different types according to the purpose for which they have been developed. These are home-gardens or kitchen-gardens, market-gardens, truck-gardens, gardens for processing, gardens for vegetable-forcing, gardens for seed production and Floating vegetable garden.

Advantages of Kitchen Garden

- It is best means of recreation and exercise.

- An excellent hobby and healthy occupation for young and old during their leisure time.

- Cut down the expenditure on purchase of vegetables.

- An ideal medium for training children in duty and order.

- Vegetables grown in kitchen garden are fresh and are free from market infection.

1. Home or Kitchen garden: is a vegetable garden where vegetable crops are grown in the back-yard of a house or any available space in the home compound to meet the daily requirement of the family. The layout of a home- garden will differ from individual to individual. However, broadly, a city home- gardener will follow a very intensive method of vegetable-growing compared with that followed by a home-gardener in a village.

2. Market garden: A market garden produces vegetables for the local market. Most of such types of gardens are located within 15 to 20 km from a city. The cropping pattern depends on the demands of the local market. The land being costly, intensive methods of cultivation are followed. A market gardener will like to grow early varieties to catch the early market. The high cost of land and labour is compensated by the availability of city compost, sludge and water near cities and high return on the produce.

3. Truck garden: A truck-garden produces selected crops in a relatively large quantity for distant markets. It generally follows a more extensive method of cultivation than the market-garden. The

commodities raised are usually sold through middlemen. The location is determined by soil and climatic factors suitable for raising particular crops. Truck gardener should be a specialized person and expert in the loge scale production and handling of some special crops. The cost of labour and land may cheap and he follows mechanized method of cultivation, hence his cost of cultivation is less. The net income is also less as this includes cost of transport and charges of middlemen.

4. Vegetable garden for processing: A vegetable garden for processing develops around the processing factories and is mainly responsible for supplying vegetables to the factories regularly. This type of garden grows particular varieties suitable for canning, dehydration or freezing. These gardens specialized in growing only a few vegetables in bulk. They choose heavier soil since their main consideration is a high and continuous yield rather than an early yield. The prices are paid on fixed contract basis on weight and quality of the product. The return may be low but the cost of marketing and the transport charges are neglisible.

5. Vegetable-forcing: Vegetable-forcing is concerned with the production of vegetables out of the normal season. The commonest forcing structures are glass and plastic houses. The vegetables commonly grown under these are tomato and cucumber. These are mostly used during wintr in the temperate region. Crops like cucumber and tomato can not be grown out side during coller months and as they are required throughout the year, they are grown under protection. Special varieties have been developed which do better under these structures. The growing of summer vegetables on river beds during winter months with the help of organic manure and wind breaks of dry grass is also a type of vegetable forcing. Sometimes for an early produce seedlings of crops like tomato or brinjal are forced to germinate in small protected structures. This may also be regarded as a type of vegetable forcing.

6. Vegetable garden for seed production: Vegetable seed production is rather a specialized type of vegetable-growing. A thorough knowledge of a vegetable crop in respect of its growth habits, mode of pollination, proper isolation distance, etc. are of prime importance in the production of quality seed. The handling of the seed-crop, its curing, threshing, cleaning, grading, packing and storage need specialized knowledge. A vegetable garden for seed production is, therefore, considered a special type of garden. The third and fourth stages of seed multiplication i.e registered and certified seeds are usually multiplied by growers. This an expanding industry in India and has good future.

7. Floating vegetable garden: Floating vegetable garden is seen on the Dal Lake of the Kashmir valley. Most of the summer vegetables are supplied to Srinagar from these gardens. Afloating base is first made from the root of Typha grass which grows wild in some parts of the lake. Once this floating base is ready, seedlings are transplanted on leaf compost made of the vegetation growing wild in the lake. All the intercultural operations and occasional sprinkling of water are done from boats.

Vegetable Crops

The term vegetables or vegetable crops refers to a classification of agricultural crops under horticulture. These crops are plants having edible parts that are used in culinary preparations either cooked or raw, as in salad recipes.

As to how many types, groupings, or classifications of vegetables are there, there can be no definitive answer. Just like food recipes, there can be too many.

The classifications can vary depending on various consideration such as the taxonomic classifications of the crops (e.g., by family or by genus), the part of plant that is edible (e.g., root vs. stem) and its stage of development (mature vs. young), their particular use in culinary preparation (for example, cooked vs. uncooked).

Classification of Vegetable Crops

There are different methods of classification of vegetables e.g:

1.   Botanical classification.

2.   Classification based on climatic zones.

3.   Classification based on the growing seasons.

4.   Classification based on economic parts used as vegetables.

5.   Classification based on method of cultivation.

Botanical Classification

Plants are divided into four great groups or sub-communities e.g:

1.   Thallophyta: Thallophytes

2.   Bryophyta: mosses

3.   Pteridophyta: ferns

4.   Spermatophyta: seed plants.

Spermatophytes are further divided into two divisions Gymnospermae and Angiospermae. Gymnosperm produces naked ovule or ovules are not enclosed in the ovary. Angiosperms produce ovules enclosed in an ovary. All vegetable crops belong to division Angiosperms of sub-community spermatophyte. Angiospermae is further divided into two classes Monocotyledonae and Dicotyledonae.

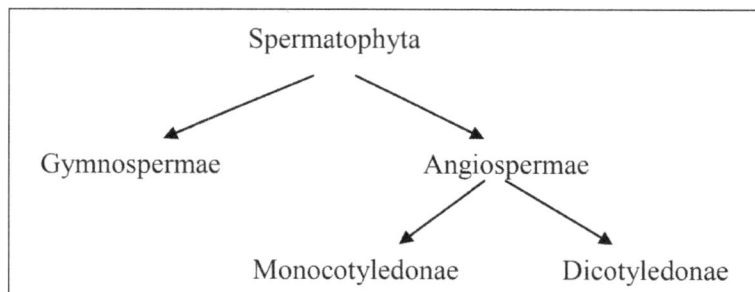

Most of the vegetables belong to the class Dicotyledonae. These classes are further divided into family, genus, species, sub-species, and finally botanical variety. The cultural operations of the vegetables belonging to the same family are not always similar e.g. potato and tomato belong to the same family but their cultural requirements are very different.

Table: Botanical classification of Some of the important Vegetables.

Common name	Family	Genus	Species
	A. Monocotyledonae		
Onion(Piyaj)	Alliaceae	Allium	cepa
Sweet corn (Makki)	Poaceae	Zea	mays
	B. Dicotyledonae		
Tomato (Tamatar)	Solanaceae	Solanum	lycopersicum
Brinjal (Baingan)		Solanum	melongena
Bell Pepper (Shimla Mirch)		Capsicum	annuum
Okra (Bhindi)	Malvaceae	Abelmoschus	esculentus
French bean (Frasbean)	Leguminosae	Phaseolus	vulgaris
Cucumber (Khira)	Cucurbitaceae	Cucmis	sativus
Bottle gourd (Ghiyya)		Lagenaria	siceraria
Bitter gourd (Karela)		Momordica	Charanita
Musk melon (Kharbuja)		Cucumis	melo
Water melon (Tarbooj)		Citrullus	lunatus
Garden pea (Matar)	Fabaceae	Pisum	sativum
Cauliflower (Phool gobhi)	Brassicaceae	Brassica	oleracea var. botrytis
Cabbage (Band gobhi)		Brassica	oleracea var. capitata
Carrot (Gajar)	Umbelliferae	Dacus	carota
Radish (Mooli)	Brassicaceae	Raphanus	sativus

Classification based on Climatic Zones

1. Tropical vegetables: Tomato, brinjal cucumber, okra, French bean, cowpea, most of cucurbits, amaranthus, cluster bean.

2. Sub-tropical vegetables: Okra, cucumber, brinjal, chilli, tomato, gourds (all), ginger, turmeric, cowpea.

3. Temperate vegetable crops: Cauliflower, cabbage, broccoli, radish, carrot, turnip, spinach, onion, garlic, pea, fenugreek, potato, asparagus and rhubarb.

Classification based on Hardiness

Also known as thermo Classification. Under this classification vegetables are grouped according to their ability to withstand frost. This class helps to know the season of cultivation of vegetables and is classified in three classes:

Hardy vegetables (withstand frost without any injury).	Broccoli, cabbage, pea, Brussels sprout, garlic, onion, leek, radish, spinach, turnip, parsley etc.
Semi-hardy vegetables (Generally they are not injured by light frost).	Carrot, cauliflower, potato, celery, lettuce, beet, palak etc.
Tender vegetables (can not withstand frost and are even killed by light frost).	Tomato, chilli, brinjal, cucumber, okra and all cucurbits, pea, French bean, sweat potato, cassava, yam drumstick, elephant foot, yam.

Classification based on Growing Season

1. Summer or warm season vegetable crops: These vegetables need optimum monthly average temperature of 20-27 °C for better growth and development. However, they can tolerate minimum temperature of 15 °C. e.g. tomato, brinjal cucumber, okra, French bean, cowpea, most of cucurbits, amaranthus, cluster bean.

2. Winter or cool season vegetable crops: Optimum monthly average temperature for better growth and development of these vegetables is 12-17 °C though can tolerate minimum temperature of 5 °C. e.g. cauliflower, cabbage, broccoli, radish, carrot, turnip, spinach, onion, garlic, pea, fenugreek, potato etc. Asparagus and Rhubarb can tolerate even temperature of 1 °C.

Classification based on Tolerance to Soil Reaction

In this classification vegetables are classified in 3 groups according to their tolerance to soil acidity:

Slightly tolerant (pH 6.0-6.0)	Moderately tolerant (6.8-5.5)	Very highly tolerant (6.8-5.0)
Broccoli, cabbage, cauliflower, bhindi, spinach, leek, Chinese cabbage, lettuce, asparagus, muskmelon, onion.	Beans, carrot, cucumber, brinjal, garlic, pea, tomato, radish, turnip, Brussels's sprouts, knolkhol, pumpkin.	Potato, sweet potato, watermelon, chicory, rhubarb.

Classification based on Salt Tolerance

Vegetables are grouped in three categories:

Sensitive	Moderately resistant	Resistant/tolerant
Pea, beans, radish, potato, brinjal, sweet potato.	Onion, carrot, cabbage, cauliflower, broccoli, tomato, melons, chilli.	Asparagus, beet, lettuce, bitter gourd, ash gourd.

Classification based on Photo Period Requirement

Vegetables are grouped according to the period for which the light is available. The response of plants to light for induction of flowering is called photo-periodism and based on it vegetables are classified in three groups:

Long day vegetables and shorter night (8 -10 hours of dark)	Short day vegetables (10- 14 hours dark)	Day neutral vegetables (Photo insensitive) (not influenced by day length)
Onion, cabbage, cauliflower, potato, radish, lettuce, knolkhol, turnip, carrot.	Sweet potato, lablab bean, winger bean, cluster bean.	Tomato, brinjal, chilli, okra, frenchbean, cucumber, cowpea.

Classification based on Rooting Depth

The knowledge of rooting depth is essential for scheduling the time and quantity of irrigation water. According to this class vegetables are classified into five categories:

Very shallow rooted (15-30 cm)	Shallow rooted (30- 60 cm)	Moderately deep rooted (60-90 cm)	Deep rooted (90-120 cm)	Very deep rooted (120-180 cm)
Onion, lettuce	Cabbage, cauliflower, garlic, celery, palak, potato, spinach, cowpea, radish, broccoli, Brussels's sprout	Brinjal, cucumber, muskmelon, frenchbean, carrot, beet	Chilli, turnip, summer squash, pea, rutabaga	Asparagus, artichoke, limabean, pumpkin, sweet potato, tomato, watermelon

- Shallow rooted require frequent and light irrigation.

- Deep rooted require less but heavy irrigation.

Classification based on Economic Parts used as Vegetables:

Leaves	Flower	Fruits	Modified stem	Under ground (plant parts)
Cabbage, palak, fenugreek, amaranthus, lettuce, celery, parsley	Broccoli, Globe artichoke	Tomato, brinjal, chilli, beans, okra, and all cucurbits	Knolkhol, cauliflower, asparagus	Carrot, turnip, beet, radish, potato, sweet potato, taro, ginger, garlic, onion, elephant foot yam, cassava

From the above classifications, the vegetables categorized in a particular group have different method of their cultivation. This means that these groups are unable to avoid repetition with respect to their method of cultivation. Therefore, there is necessity to classify the vegetables in such groups where they have almost similar cultivation techniques.

Classification based on Methods of Culture

This is the most convenient method of classification. In this classification, vegetable crops having same cultural requirements are placed together. As a consequence, it is possible to give the general cultural practices for the group without the necessity of repetition while describing the individual crop. Some groups like cucurbits, cole crops, solanaceous and bulb crops not only have similar cultural requirements for the group but the crops belonging to each group also have the same family.

Most of the crops belonging to bulb and salad group have similar temperature requirements. Therefore, this method of classification even though not in all but in the majority of cases fulfills the basic requirements of classification of vegetables:

- Group 1: Potato

- Group 2: Solanaceous fruits e.g. Tomato, brinjal, capsicum, chilli etc.

- Group 3: Cole crops e.g. cabbage, cauliflower, broccoli, knolkhol, kale.

- Group 4: Cucurbits e.g. cucumber, bottle gourd, bitter gourd, ridge gourd, snake gourd, water melon, pumpkin, summer squash and winter squash.

- Group 5: Root crops e.g. Radish, carrot, turnip, beat.

- Group 6: Bulb crops e.g. Onion, garlic, and leek.

- Group 7: Salad crops e.g. Lettuce, celery, parsley.

- Group 8: Greens and pot herbs e.g. Spinach, coriander, fenugreek, palak, beat, leak, amaranthus. Group 9: Peas and beans e.g. Pea, Frenchbean, asparagus bean, lima beans, cluster bean, cowpea etc.

- Group 10: Tuber crops other than potato e.g. Taro, yarn, elephant foot yam.

- Group 11: Sweet potato.

- Group 12: Okra.

- Group 13: Pointed gourd.

- Group 14: Temperate perennials e.g. Globe artichoke, Rhubarb.

- Group 15: Tropical perennials vegetables e.g. Curry leaves, drum stick.

- Group 16: Chow-chow (Chayote).

Potato

Potato (Solanum tuberosum) is the most important food crop of the world Potato is grown in more than 100 countries, under temperate, subtropical and tropical conditions. It is essentially a "cool weather crop", with temperature being the main limiting factor on production: tuber growth is sharply inhibited in temperatures below 10 °C (50 °F) and above 30 °C (86 °F), while optimum yields are obtained where mean daily temperatures are in the 18 to 20 °C (64 to 68 °F) range.

For that reason, potato is planted in early spring in temperate zones and late winter in warmer regions, and grown during the coolest months of the year in hot tropical climates. In some sub-tropical highlands, mild temperatures and high solar radiation allow farmers to grow potatoes throughout the year, and harvest tubers within 90 days of planting (in temperate climates, such in northern Europe, it can take up to 150 days).

The potato is a very accommodating and adaptable plant, and will produce well without ideal soil and growing conditions. However, it is also subject to number of pests and diseases. To prevent the build-up of pathogens in the soil, farmers avoid growing potato on the same land from year to year. Instead, they grow potato in rotations of three or more years, alternating with other, dissimilar crops, such as maize, beans and alfalfa. Crops susceptible to the same pathogens as potato (e.g. tomato) are avoided in order to break potato pests' development cycle.

With good agricultural practices, including irrigation when necessary, a hectare of potato in the temperate climates of northern Europe and North America can yield more than 40 tonnes of fresh tubers within four months of planting. In most developing countries, however, average yields are much lower - ranging from as little as five tonnes to 25 tonnes - owing to lack of high quality seed and improved cultivars, lower rates of fertilizer use and irrigation, and pest and disease problems.

Soil and Land Preparation

The potato can be grown almost on any type of soil, except saline and alkaline soils. Naturally loose soils, which offer the least resistance to enlargement of the tubers, are preferred, and loamy and sandy loam soils that are rich in organic matter, with good drainage and aeration, are the most suitable. Soil with a pH range of 5.2-6.4 is considered ideal.

Growing potatoes involves extensive ground preparation. The soil needs to be harrowed until completely free of weed roots. In most cases, three ploughings, along with frequent harrowing and rolling, are needed before the soil reaches a suitable condition: soft, well-drained and well-aerated.

Planting

The potato crop is usually grown not from seed but from "seed potatoes" - small tubers or pieces of tuber sown to a depth of 5 to 10 cm. Purity of the cultivars and healthy seed tubers are essential for a successful crop. Tuber seed should be disease-free, well-sprouted and from 30 to 40 g each in weight. Use of good quality commercial seed can increase yields by 30 to 50 percent, compared to farmers' own seed, but expected profits must offset the higher cost.

The planting density of a row of potatoes depends on the size of the tubers chosen, while the inter-row spacing must allow for ridging of the crop. Usually, about two tonnes of seed potatoes are sown per hectare. For rainfed production in dry areas, planting on flat soil gives higher yields (thanks to better soil water conservation), while irrigated crops are mainly grown on ridges.

Stages in crop development:

- Planted seed tuber
- Vegetative growth
- Tuber initiation
- Tuber bulking.

Crop Care

During the development of the potato canopy, which takes about four weeks, weeds must be controlled in order to give the crop a "competitive advantage". If the weeds are large, they must be removed before ridging operations begin. Ridging (or "earthing up") consists of mounding the soil from between the rows around the main stem of the potato plant. Ridging keeps the plants upright and the soil loose, prevents insect pests such a tuber moth from reaching the tubers; and helps prevent the growth of weeds.

After earthing up, weeds between the growing plants and at the top of the ridge are removed mechanically or using herbicides. Ridging should be done two or three times at an interval of 15 to 20 days. The first should be done when the plants are about 15-25 cm high; the second is often done to cover the growing tubers.

Manuring and Fertilization

The use of chemical fertilizer depends on the level of available soil nutrients - volcanic soils, for example, are typically deficient in phosphorus - and in irrigated commercial production, fertilizer requirements are relatively high. However, potato can benefit from application of organic manure at the start of a new rotation - it provides a good nutrient balance and maintains the structure to the soil. Crop fertilization requirements need to be correctly estimated according to the expected yield, the potential of the variety and the intended use of the harvested crop.

Water Supply

The soil moisture content must be maintained at a relatively high level. For best yields, a 120 to 150 day crop requires from 500 to 700 mm (20 to 27.5 inches) of water. In general, water deficits in the middle to late part of the growing period tend to reduce yield more than those in the early part. Where supply is limited, water is directed towards maximizing yield per hectare rather than being applied over a larger area.

Because the potato has a shallow root system, yield response to frequent irrigation is considerable, and very high yields are obtained with mechanized sprinkler systems that replenish evapotranspiration losses every one or two days. Under irrigation in temperate and subtropical climates, a crop of about 120 days can produce yields of 25 to 35 tonnes/ha (11 to 15.6 tons per acre), falling to 15 to 25 tonnes/ha (6.6 to 15.6 tons per acre) in tropical areas.

Pests and Diseases

Against diseases, a few basic precautions – crop rotation, using tolerant varieties and healthy, certified seed tubers - can help avoid great losses. There is no chemical control for bacterial and viral diseases but they can be controlled by regular monitoring (and when necessary, spraying) of their aphid vectors. The severity of fungal diseases such as late blight depends, after the first infection, mainly on the weather - persistence of favourable conditions, without chemical spraying, can quickly spread the disease.

Insect pests can wreak havoc in a potato patch. Recommended control measures include regular monitoring and steps to protect the pests' natural enemies. Even damage caused by the Colorado Potato Beetle, a major pest, can be reduced by destroying beetles, eggs and larvae that appear early in the season, while sanitation, crop rotations and use of resistant potato varieties help prevent the spread of nematodes.

Harvesting

Yellowing of the potato plant's leaves and easy separation of the tubers from their stolons indicate that the crop has reached maturity. If the potatoes are to be stored rather than consumed immediately, they are left in the soil to allow their skins to thicken - thick skins prevent storage diseases and shrinkage due to water loss. However, leaving tubers for too long in the ground increases their exposure to a fungal incrustation called black scurf.

To facilitate harvesting, the potato vines should be removed two weeks before the potatoes are dug up. Depending on the scale of production, potatoes are harvested using a spading fork, a plough or commercial potato harvesters that unearth the plant and shake or blow the soil from the tubers. During harvesting, it is important to avoid bruising or other injury, which provide entry points for storage diseases.

Storage

Since the newly harvested tubers are living tissue – and therefore subject to deterioration - proper storage is essential, both to prevent post-harvest losses of potatoes destined for fresh consumption or processing, and to guarantee an adequate supply of seed tubers for the next cropping season.

For ware and processing potatoes, storage aims at preventing "greening" (the build up of chlorophyll beneath the peel, which is associated with solanine, a potentially toxic alkaloid) and losses in weight and quality. The tubers should be kept at a temperature of 6 to 8 °C degrees, in a dark, well-ventilated environment with high relative humidity (85 to 90 percent). Seed tubers are stored, instead, under diffused light in order to maintain their germination capacity and encourage development of vigorous sprouts. In regions, such as northern Europe, with only one cropping season and where storage of tubers from one season to the next is difficult without the use of costly refrigeration, off-season planting may offer a solution.

Tomato

Tomato (Lycopersicon esculentum) is the second most important vegetable crop next to potato. Present world production is about 100 million tons fresh fruit from 3.7 million ha.

Tomato is a rapidly growing crop with a growing period of 90 to 150 days. It is a daylength neutral plant. Optimum mean daily temperature for growth is 18 to 25 °C with night temperatures between 10 and 20 °C. Larger differences between day and night temperatures, however, adversely affect yield. The crop is very sensitive to frost. Temperatures above 25 °C, when accompanied by high humidity and strong wind, result in reduced yield. Night temperatures above 20 °C accompanied by high humidity and low sunshine lead to excessive vegetative growth and poor fruit production. High humidity leads to a greater incidence of pests and diseases and fruit rotting. Dry climates are therefore preferred for tomato production.

Tomato can be grown on a wide range of soils but a well-drained, light loam soil with pH of 5 to 7 is preferred. Waterlogging increases the incidence of diseases such as bacterial wilt. The fertilizer requirements amount, for high producing varieties, to 100 to 150 kg/ha N, 65 to 110 kg/ha P and 160 to 240 kg/ha K.

The seed is generally sown in nursery plots and emergence is within 10 days. Seedlings are transplanted in the field after 25 to 35 days. In the nursery the row distance is about 10 cm. In the field spacing ranges from 0.3/0.6 x 0.6/1 m with a population of about 40,000 plants per ha. The crop should be grown in a rotation with crops such as maize, cabbage, cowpea, to reduce pests and disease infestations.

The crop is moderately sensitive to soil salinity. Yield decrease at various ECe values is: 0% at ECe 2.5 mmhos/cm, 10% at 3.5, 25% at 5.0, 50% at 7.6 and 100%. at ECe 12.5 mm hos/cm. The most sensitive period to salinity is during germination and early plant development, and necessary leaching of salts is therefore frequently practised during pre-irrigation or by over-watering during the initial irrigation application.

The graph below depicts the crop stages of tomato, and the table summarises the main crop coefficients used for water management.

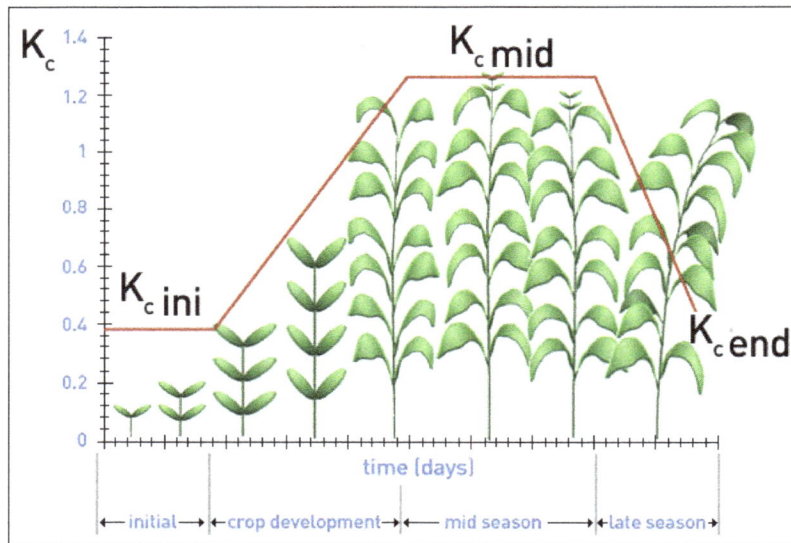

Crop characteristic	Stages of Development					Plant date	Region
	Initial	Crop Development	Mid-season	Late	Total		
Stage length, days	30	40	40	25	135	Jan	Arid Region
	35	40	50	30	155	Apr/May	Calif., USA
	25	40	60	30	155	Jan	Calif.,
	35	45	70	30	180	Oct/Nov	Desert USA
	30	40	45	30	145	Apr/May	Arid Region Mediterranean
Depletion Coefficient, p:	0.3	>>	0.4	0.5	0.3		
Root Depth, m	0.25	>>	-	1.0	-		
Crop Coefficient, Kc	0.6	>>	1.15	0.7-0.9	-		
Yield Response Factor, Ky	0.4	1.1	0.8	0.4	1.05		

Water Requirements

Total water requirements (ETm) after transplanting, of a tomato crop grown in the field for 90 to 120 days, are 400 to 600 mm, depending on the climate. Water requirements related to reference evapotranspiration (ETo) in mm/period are given by the crop factor (Kc) for different crop development stages, or: during the initial stage 0. 4-0. 5 (10 to 15 days), the development stage 0. 7-0.8 (20 to 30 days), the mid- season stage 1.05-1.25 (30 to 40 days), the late-season stage 0.8-0.9 (30 to 40 days) and at harvest 0.6-0.65.

Water Supply and Crop Yield

The relationships between relative yield decrease (1 - Ya/Ym) and relative evapotranspiration deficit for the total growing period are shown in the figure below.

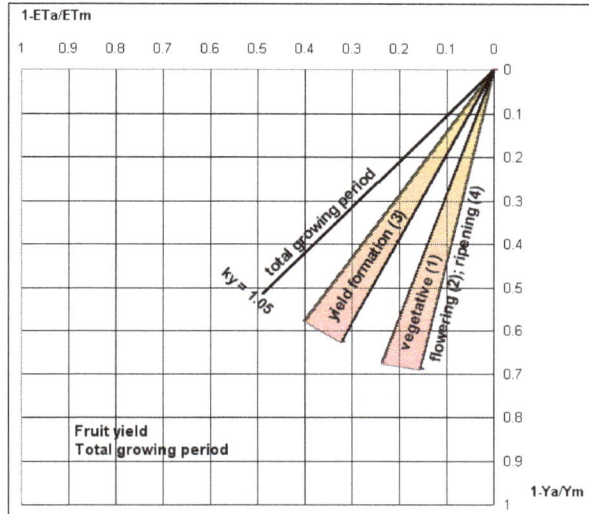

The relationships between relative yield decrease (1 - Ya/Ym) and relative evapotranspiration deficit for the individual growing periods are shown in the figure below.

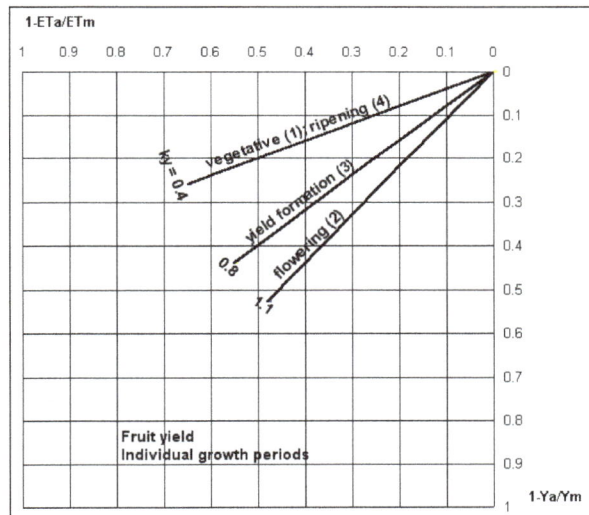

The plant produces flowers from bottom to top during the active development of the stem. Fruits can be harvested while the plant is still flowering at the top. Some-times three flowering periods related to three harvests can be distinguished. However, for mechanical harvesting where the fruits are used for tomato paste, only one picking is made. Water supply needs to be adjusted according to the use of the product, c. g. for salad or paste.

Highest yields of salad tomatoes are obtained by frequent, light irrigation. Where mechanical harvesting is used, heavy, infrequent irrigation is more appropriate with the last irrigation applied long before harvest.

Following table presents the growth periods of tomato at the first harvest.

Stage	Development Stage	Stage length, days
0	Establishment	25-35
1	Vegetative	20-25
2	Flowering	20-30
3	Yield formation	20-30
4	Ripening	15-20
Total		100-140 days

For subsequent harvest periods, 2, 3 and 4 will overlap and an additional 20 to 30 days are required for each harvest.

The relationship between relative yield decrease (1 - Ya/Ym) and relative evapotranspiration deficit (1 - ETa/ETm) is given in the following figure. The crop is most sensitive to water deficit during and immediately after transplanting and during, flowering and yield formation. Water deficit during the flowering period causes flower drop. Moderate water deficit during the vegetative period enhances root growth.

For high yield and good quality, the crop needs a controlled supply of water throughout the growing period. Whereas under water limiting conditions some water savings may be made during the vegetative and ripening periods, water supply should preferably be directed toward maximizing production per ha rattler than extending the cultivated area under limited water supply.

Water Uptake

The crop has a fairly deep root system and in deep soils roots penetrate up to some 1. 5 m. The maximum rooting depth is reached about 60 days after transplanting. Over 80 percent of the total water uptake occurs in the first 0.5 to 0.7 m and 100 per-cent of the water uptake of a full grown crop occurs from the first 0.7 to 1.5 m (D = 0.7 - 1.5 m). Under conditions when maximum evapotranspiration (ETm) is 5 to 6 mm/ day water uptake to meet full crop water requirements is affected when more than 40 percent of the total available soil water has been depleted (p = 0.4).

Irrigation Scheduling

The crop performance is sensitive to the irrigation practices. In general a prolonged severe water deficit limits growth and reduces yields which cannot be corrected by heavy watering later on. Highest demand for water is during flowering. However, with holding irrigation during this period is sometimes recommended to force less mature plants into flowering in order to obtain uniform flowering and ripening. Care should be exercised in this to avoid damage to the mature plants.

Excessive watering during the flowering period has been shown to increase flower drop and reduce fruit set. Also this may cause excessive vegetative growth and a delay in ripening. Water supply during and after fruit set must be limited to a rate which will prevent stimulation of new growth at the expense of fruit development. Heavy, irregular irrigations or dry periods alternating with wet periods should be avoided. For production of salad tomato with more than one harvest, the crop flourishes best under light, frequent irrigation, well-distributed over the growing period with the soil depletion level during the different growth periods remaining below 40 percent ($p < 0.4$). This promotes optimum growth during the total growing period and results in high yield of good quality. With one harvest uniform ripening is required and the depletion level during this period may increase to 60 to 70 percent.

When water supply is limited, application for a salad crop can be concentrated during periods of transplanting, flowering and yield formation. For a crop grown for paste production, a more extensive irrigation may be applied with last heavy irrigation applied prior to flowering.

Irrigation Methods

Surface irrigation by furrow is commonly practised. Under sprinkler irrigation the occurrence of fungal diseases and possibly bacterial canker may become a major problem. Further, under sprinkler, fruit set may be reduced with an increase in fruit rotting. In the case of poor quality water, leaf burn will occur with sprinkler irrigation; this may be reduced by sprinkling at night and shifting of sprinkler lines with the direction of the prevailing wind. Due to the crops specific demands for a high soil water content achieved without leaf wetting, trickle or drip irrigation has been successfully applied.

Yield

Frequent light irrigation improve the size, shape, juiciness and colour of the fruit, but total solids (dry matter content) and acid content will be reduced. However, the decrease in solids will lower the fruit quality for processing. In selecting the irrigation practices consideration must therefore be given to teh type of end product required. Prolonged water deficits leads to fruit cracking. Where fruit rot is a problem, frequent sprinkler irrigation should be avoided during the period of yield formation.

A good commercial yield under irrigation is 45 to 65 tons/ha fresh fruit, of which 80 to 90 percent is moisture, depending on the use of the product. the water utilization efficiency for harvested yield (Ey) for fresh tomatoes is 10 to 12 kg/m^3.

Carrot

The carrot (Daucus carota subsp. sativus) is a root vegetable, usually orange in colour, though purple, black, red, white, and yellow cultivars exist. Carrots are a domesticated form of the wild carrot, Daucus carota, native to Europe and southwestern Asia. The plant probably originated in Persia and was originally cultivated for its leaves and seeds. The most commonly eaten part of the plant is the taproot, although the stems and leaves are eaten as well. The domestic carrot has been selectively bred for its greatly enlarged, more palatable, less woody-textured taproot.

The carrot is a biennial plant in the umbellifer family Apiaceae. At first, it grows a rosette of leaves while building up the enlarged taproot. Fast-growing cultivars mature within three months (90 days) of sowing the seed, while slower-maturing cultivars need a month longer (120 days). The roots contain high quantities of alpha- and beta-carotene, and are a good source of vitamin K and vitamin B6, but the belief that eating carrots improves night vision is a myth put forward by the British in World War II to mislead the enemy about their military capabilities.

The United Nations Food and Agriculture Organization (FAO) reports that world production of carrots and turnips (these plants are combined by the FAO) for the calendar year 2013 was 37.2 million tonnes; almost half (~45%) were grown in China. Carrots are widely used in many cuisines, especially in the preparation of salads, and carrot salads are a tradition in many regional cuisines.

Daucus carota is a biennial plant. In the first year, its rosette of leaves produces large amounts of sugars, which are stored in the taproot to provide energy for the plant to flower in the second year.

Seedlings shortly after germination.

Soon after germination, carrot seedlings show a distinct demarcation between taproot and stem: the stem is thicker and lacks lateral roots. At the upper end of the stem is the seed leaf. The first true leaf appears about 10–15 days after germination. Subsequent leaves are alternate (with a single leaf attached to a node), spirally arranged, and pinnately compound, with leaf bases sheathing the stem. As the plant grows, the bases of the seed leaves, near the taproot, are pushed apart. The stem, located just above the ground, is compressed and the internodes are not distinct. When the seed stalk elongates for flowering, the tip of the stem narrows and becomes pointed, and the stem extends upward to become a highly branched inflorescence up to 60–200 cm (20–80 in) tall.

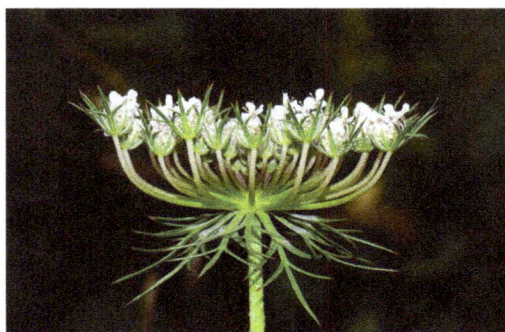

Daucus carota umbel (inflorescence). Individual flowers are borne on undivided pedicels originating from a common node.

Most of the taproot consists of a pulpy outer cortex (phloem) and an inner core (xylem). High-quality carrots have a large proportion of cortex compared to core. Although a completely xylem-free carrot is not possible, some cultivars have small and deeply pigmented cores; the taproot can appear to lack a core when the colour of the cortex and core are similar in intensity. Taproots are typically long and conical, although cylindrical and nearly-spherical cultivars are available. The root diameter can range from 1 cm (0.4 in) to as much as 10 cm (4 in) at the widest part. The root length ranges from 5 to 50 cm (2 to 20 in), although most are between 10 and 25 cm (4 and 10 in).

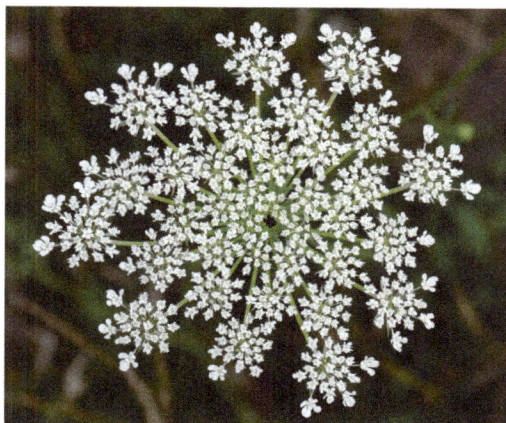

Top view of Daucus carota inflorescence, showing umbellets; the central flower is dark red.

Flower development begins when the flat meristem changes from producing leaves to an uplifted, conical meristem capable of producing stem elongation and a cluster of flowers. The cluster is a compound umbel, and each umbel contains several smaller umbels (umbellets). The first (primary) umbel occurs at the end of the main floral stem; smaller secondary umbels grow from the main branch, and these further branch into third, fourth, and even later-flowering umbels. A large, primary umbel can contain up to 50 umbellets, each of which may have as many as 50 flowers; subsequent umbels have fewer flowers. Individual flowers are small and white, sometimes with a light green or yellow tint. They consist of five petals, five stamens, and an entire calyx. The stamens usually split and fall off before the stigma becomes receptive to receive pollen. The stamens of the brown, male, sterile flowers degenerate and shrivel before the flower fully opens. In the other type of male sterile flower, the stamens are replaced by petals, and these petals do not fall off. A nectar-containing disc is present on the upper surface of the carpels.

Flowers consist of five petals, five stamens, and an entire calyx.

Flowers change sex in their development, so the stamens release their pollen before the stigma of the same flower is receptive. The arrangement is centripetal, meaning the oldest flowers are near the edge and the youngest flowers are in the center. Flowers usually first open at the outer edge of the primary umbel, followed about a week later on the secondary umbels, and then in subsequent weeks in higher-order umbels. The usual flowering period of individual umbels is 7 to 10 days, so a plant can be in the process of flowering for 30–50 days. The distinctive umbels and floral nectaries attract pollinating insects. After fertilization and as seeds develop, the outer umbellets of an umbel bend inward causing the umbel shape to change from slightly convex or fairly flat to concave, and when cupped it resembles a bird's nest.

The fruit that develops is a schizocarp consisting of two mericarps; each mericarp is a true seed. The paired mericarps are easily separated when they are dry. Premature separation (shattering) before harvest is undesirable because it can result in seed loss. Mature seeds are flattened on the commissural side that faced the septum of the ovary. The flattened side has five longitudinal ribs. The bristly hairs that protrude from some ribs are usually removed by abrasion during milling and cleaning. Seeds also contain oil ducts and canals. Seeds vary somewhat in size, ranging from less than 500 to more than 1000 seeds per gram.

The carrot is a diploid species, and has nine relatively short, uniform-length chromosomes ($2n=18$). The genome size is estimated to be 473 mega base pairs, which is four times larger than *Arabidopsis thaliana*, one-fifth the size of the maize genome, and about the same size as the rice genome.

Cultivation

Carrots are grown from seed and can take up to four months (120 days) to mature, but most cultivars mature within 70 to 80 days under the right conditions. They grow best in full sun but tolerate some shade. The optimum temperature is 16 to 21 °C (61 to 70 °F). The ideal soil is deep, loose and well-drained, sandy or loamy, with a pH of 6.3 to 6.8. Fertilizer should be applied according to soil type because the crop requires low levels of nitrogen, moderate phosphate and high potash. Rich or rocky soils should be avoided, as these will cause the roots to become hairy and/or misshapen. Irrigation is applied when needed to keep the soil moist. After sprouting, the crop is eventually thinned to a spacing of 8 to 10 cm (3 to 4 in) and weeded to prevent competition beneath the soil.

Workers harvesting carrots, Imperial Valley.

Cultivation Problems

There are several diseases that can reduce the yield and market value of carrots. The most devastating carrot disease is *Alternaria* leaf blight, which has been known to eradicate entire crops. A bacterial leaf blight caused by *Xanthomonas campestris* can also be destructive in warm, humid areas. Root knot nematodes (*Meloidogyne* species) can cause stubby or forked roots, or galls. Cavity spot, caused by the oomycetes *Pythium violae* and *Pythium sulcatum*, results in irregularly shaped, depressed lesions on the taproots.

Physical damage can also reduce the value of carrot crops. The two main forms of damage are splitting, whereby a longitudinal crack develops during growth that can be a few centimetres to the entire length of the root, and breaking, which occurs postharvest. These disorders can affect over 30% of commercial crops. Factors associated with high levels of splitting include wide plant spacing, early sowing, lengthy growth durations, and genotype.

Companion Planting

Carrots benefit from strongly scented companion plants. The pungent odour of onions, leeks and chives help repel the carrot root fly, and other vegetables that team well with carrots include lettuce, tomatoes and radishes, as well as the herbs rosemary and sage. Carrots thrive in the presence of caraway, coriander, chamomile, marigold and Swan River daisy. They can also be good companions for other plants; if left to flower, the carrot, like any umbellifer, attracts predatory wasps that kill many garden pests.

Cultivars

Carrot cultivars can be grouped into two broad classes, eastern carrots and western carrots. A number of novelty cultivars have been bred for particular characteristics.

"Eastern" (a European and American continent reference) carrots were domesticated in Persia (probably in the lands of modern-day Iran and Afghanistan within West Asia) during the 10th century, or possibly earlier. Specimens of the "eastern" carrot that survive to the present day are commonly purple or yellow, and often have branched roots. The purple colour common in these carrots comes from anthocyanin pigments.

Carrot seeds

Seeds of *Daucus carota subsp. maximus* - MHNT

The western carrot emerged in the Netherlands in the 17th century. There is a popular belief that its orange colour making it popular in those countries as an emblem of the House of Orange and the struggle for Dutch independence, although there is little evidence for this. The orange colour results from abundant carotenes in these cultivars.

Western carrot cultivars are commonly classified by their root shape. The four general types are:

- Chantenay Carrots: Although the roots are shorter than other cultivars, they have vigorous foliage and greater girth, being broad in the shoulders and tapering towards a blunt, rounded tip. They store well, have a pale-coloured core and are mostly used for processing. Cultivars include 'Carson Hybrid' and 'Red Cored Chantenay'.

- Danvers Carrots: These have strong foliage and the roots are longer than Chantaney types, and they have a conical shape with a well-defined shoulder, tapering to a point. They are somewhat shorter than Imperator cultivars, but more tolerant of heavy soil conditions. Danvers cultivars store well and are used both fresh and for processing. They were developed in 1871 in Danvers, Massachusetts. Cultivars include 'Danvers Half Long' and 'Danvers 126'.

- Imperator Carrots: This cultivar has vigorous foliage, is of high sugar content, and has long and slender roots, tapering to a pointed tip. Imperator types are the most widely cultivated by commercial growers. Cultivars include 'Imperator 58' and 'Sugarsnax Hybrid'.

- Nantes Carrots: These have sparse foliage, are cylindrical, short with a more blunt tip than Imperator types, and attain high yields in a range of conditions. The skin is easily damaged and the core is deeply pigmented. They are brittle, high in sugar and store less well than other types. Cultivars include 'Nelson Hybrid', 'Scarlet Nantes' and 'Sweetness Hybrid'.

One particular cultivar lacks the usual orange pigment due to carotene, owing its white colour to a recessive gene for tocopherol (vitamin E), but this cultivar and wild carrots do not provide nutritionally significant amounts of vitamin E.

Production

Production of carrots (and turnips) in 2016	
Country	Production (millions of tonnes)
China	20.5
European Union	5.9
Uzbekistan	2.3
Russia	1.8
United States	1.4
Ukraine	0.9
World	42.7

In 2016, world production of carrots (combined with turnips) was 42.7 million tonnes, with China producing 48% of the world total (20.5 million tonnes, table). Other major producers were the European Union, Uzbekistan, Russia, the United States, and Ukraine.

Storage

Carrots can be stored for several months in the refrigerator or over winter in a moist, cool place. For long term storage, unwashed carrots can be placed in a bucket between layers of sand, a 50/50 mix of sand and wood shavings, or in soil. A temperature range of 32 to 40 °F (0 to 5 °C) is best.

Consumption

Carrots can be eaten in a variety of ways. Only 3 percent of the β-carotene in raw carrots is released during digestion: this can be improved to 39% by pulping, cooking and adding cooking oil. Alternatively they may be chopped and boiled, fried or steamed, and cooked in soups and stews, as well as baby and pet foods. A well-known dish is carrots julienne. Together with onion and celery, carrots are one of the primary vegetables used in a mirepoix to make various broths.

The greens are edible as a leaf vegetable, but are rarely eaten by humans; some sources suggest that the greens contain toxic alkaloids. When used for this purpose, they are harvested young in high-density plantings, before significant root development, and typically used stir-fried, or in salads. Some people are allergic to carrots. In a 2010 study on the prevalence of food allergies in Europe, 3.6 percent of young adults showed some degree of sensitivity to carrots. Because the major carrot allergen, the protein Dauc c 1.0104, is cross-reactive with homologues in birch pollen (Bet v 1) and mugwort pollen (Art v 1), most carrot allergy sufferers are also allergic to pollen from these plants.

In India carrots are used in a variety of ways, as salads or as vegetables added to spicy rice or dal dishes. A popular variation in north India is the Gajar Ka Halwa carrot dessert, which has carrots grated and cooked in milk until the whole mixture is solid, after which nuts and butter are added. Carrot salads are usually made with grated carrots with a seasoning of mustard seeds and green chillies popped in hot oil. Carrots can also be cut in thin strips and added to rice, can form part of a dish of mixed roast vegetables or can be blended with tamarind to make chutney.

Since the late 1980s, baby carrots or mini-carrots (carrots that have been peeled and cut into uni-form cylinders) have been a popular ready-to-eat snack food available in many supermarkets. Car-rots are puréed and used as baby food, dehydrated to make chips, flakes, and powder, and thinly sliced and deep-fried, like potato chips.

The sweetness of carrots allows the vegetable to be used in some fruit-like roles. Grated carrots are used in carrot cakes, as well as carrot puddings, an English dish thought to have originated in the early 19th century. Carrots can also be used alone or blended with fruits in jams and preserves. Carrot juice is also widely marketed, especially as a health drink, either stand-alone or blended with juices extracted from fruits and other vegetables.

Highly excessive consumption over a period of time results in a condition of carotenemia which is a yellowing of the skin caused by a build up of carotenoids.

Nutrition

Raw carrots are 88% water, 9% carbohydrates, 0.9% protein, 2.8% dietary fiber, 1% ash and 0.2% fat. Carrot dietary fiber comprises mostly cellulose, with smaller proportions of hemicellulose, lig-nin and starch. Free sugars in carrot include sucrose, glucose, and fructose.

The carrot gets its characteristic, bright orange colour from β-carotene, and lesser amounts of α-carotene, γ-carotene, lutein, and zeaxanthin. α- and β-carotenes are partly metabolized into vitamin A, providing more than 100% of the Daily Value (DV) per 100 g serving of carrots. Carrots are also a good source of vitamin K (13% DV) and vitamin B6 (11% DV), but otherwise have modest content of other essential nutrients.

Night Vision

The provitamin A beta-carotene from carrots does not actually help people to see in the dark un-less they suffer from vitamin A deficiency. This myth was propaganda used by the Royal Air Force during the Second World War to explain why their pilots had improved success during night air battles, but was actually used to disguise advances in radar technology and the use of red lights on instrument panels. Nevertheless, the consumption of carrots was advocated in Britain at the time as part of a Dig for Victory campaign. A radio programme called *The Kitchen Front* encour-aged people to grow, store and use carrots in various novel ways, including making carrot jam and Woolton pie, named after the Lord Woolton, the Minister for Food. The British public during WWII generally believed that eating carrots would help them see better at night and in 1942 there was a 100,000 ton surplus of carrots from the extra production.

Onion

Onion belongs to the family Amaryllidaceae with botanical name Allium cepa. In the world total area under cultivation of onion is about 19,77,000 hectare which gives 2,79,18,000 mt. Major pro-ducing countries are North America, Japan, Spain, Netherland, Canada, etc.

Climate and Soil

An ideal soil should have pH in between 6.5 to 8. The soil should be well aerated. Heavy soil should be avoided. It is grown under a wide range of climatic conditions. However, it cannot stand too hot or too cold weather. It prefers moderate temperature in summer as well as in winter. Short days are very favorable for the formation of bulbs. It can be grown well at elevations of 1000 to 1300 m above MSL. Onion requires well drained loamy soils, rich in humus, with fairly good content of potash. The crop raised on sandy or loose soil does soils, the bulbs produced are deformed and during harvesting, many bulbs are broken and bruised and so they do not keep well in storage.

Varieties

Pusa Red, Pusa Ratnar, Pusa White Round, Patna Red, Poona Red, Arka Pragati, Arka Niketa, Patna White, Bombay White, Nasik Red.

Planting Habits

It is grow as an annual or biennial. The smooth, glaucous scope grows from 1 to 3 feet high. The scope or stem is hollow and swollen above the base. The leaves are also swollen at base and hollow. Flowers, terminal, umbel, white or faint blue, numerous, develop bulbels. Underground bulb undeveloped like the stem. Propagation by Bulbels.

Propagation

Onion belong to bulb vegetables group. Onion seeds are sown in nursery from October to November. One hectare needs 8-10 kg seeds. Seedlings are transplanted in December and January. Early transplanting yields more. Bulb and bulb-lets are also sown but needs 1000-1200 kg./hectare.

Planting

The seed of onion in the nursery from middle of October to the end of November. In the hills, the seed is sown from March to June. Planting distance for Onion (seedlings) is 15 × 10 or 20 × 10 cm.

Application of Manures and Fertilizer

Onion F.Y.M. 200-250 g/hectare

45-65 Kg N/hectare

40-60 kg P/hectare

60-100 kg K/hectare.

Irrigation

Irrigation during growth should be steady and uninterrupted otherwise dryness may cause splitting of the outer scales. Irrigation is stopped when the tops mature and start falling. Use F1 hybrids for higher yield.

Plant Protection

Pest

- Thrips- Spray Malathion or Nuvacron (0.1%).

- Borer- Spray Endosulfan (0.1%).

- Maggot- Apply Thimet 10G to soil and spray Malathion (0.05%).

Diseases

- Downy mildew- Spray Difolatan (0.1%) or Dithane M-45 (0.2%).

- Smut - Spray Captan, Biltox or Thiram 75%.

Yield

Onion gives a yield of 25 to 30 tones/hectare and garlic about 6 to 10 tonnes. Bulbs should be thoroughly cured before storage.

Storage

In case of onion crop there are various factors affecting the storing conditions. Some important are listed below.

1. Selection of Variety.
2. Fertilizers and water management.
3. Drying of onion.
4. Actual storage conditions.
5. Construction of storage structure.

1. Selection of Variety: In majority of the cases onion crop grown in rabi season are stored. This can be stored for about 4 to 5 months. In case of storage selection of variety is an important aspect where N 2-4-1, Agrifound light red, Arka niketan were found to be suitable one giving less percentage of losses.

2. Fertilizers and water management: Doses of fertilizers, type of fertilizers and irrigation affects the storing capacity of onion/shelf life of onion. Organic matter increases the storing capacity of onion. Rate of fertilizer suggested is 150:50:50, N:P:K kg per hectare respectably. In case of irrigation according to soil tag crop should be irrigated.

3. Drying of onion: Immediately after harvest let the onion should get dried for about 8 to 10 days which helps to remove the excess heat giving better colour to onion.

4. Actual storage conditions: In case of storage condition temperature and humidity are the important parameters. Optimum temperature ranges from 25 to 30 0C and RH 65 to 70%. Excess humidity and temperature results in fungal growth. On the contrary decrease in humidity

increases respiration rate causing weight loss. In May-June temperature is more and relative humidity is less which affects the weight loss. While in case of July-September temperature goes down resulting in sprouting of onion. To avoid these problems construction of proper storage structure is required.

5. Construction of storage structure:

- Storage structure should be constructed 30cm from the ground level giving cement base which avoids contact with soil moisture.

- For roofing purpose sugarcane trash, asbestos sheet is used which lowers down the temperature.

- For good aeration central height should be optimum (120cm) giving sloppy.

Structure

Care should be taken that the stored onions should not get directly expose to sunlight which gives good colour and improves the structure of onion. 1ft3 area can store 20kg of onion:

- Before and after storage stabilization of the storage structure can be done by spraying 10g.

- Bavistin + 15ml neuocron in 10 liters of water which helps in controlling the pest.

Vegetable Farming

Vegetable farming is the growing of vegetable crops, primarily for use as human food. The term vegetable in its broadest sense refers to any kind of plant life or plant product; in the narrower sense, however, it refers to the fresh, edible portion of a herbaceous plant consumed in either raw or cooked form. The edible portion may be a root, such as rutabaga, beet, carrot, and sweet potato; a tuber or storage stem, such as potato and taro; the stem, as in asparagus and kohlrabi; a bud, such as brussels sprouts; a bulb, such as onion and garlic; a petiole or leafstalk, such as celery and rhubarb; a leaf, such as cabbage, lettuce, parsley, spinach, and chive; an immature flower, such as cauliflower, broccoli, and artichoke; a seed, such as pea and lima bean; the immature fruit, such as eggplant, cucumber, and sweet corn (maize); or the mature fruit, such as tomato and pepper.

The popular distinction between vegetable and fruit is difficult to uphold. In general, those plants or plant parts that are usually consumed with the main course of a meal are popularly regarded as vegetables, while those mainly used as desserts are considered fruits. Thus, cucumber and tomato, botanically fruits, since they are the portion of the plant containing seeds, are commonly regarded as vegetables.

Types of Production

Vegetable production operations range from small patches of crops, producing a few vegetables for family use or marketing, to the great, highly organized and mechanized farms common in the most technologically advanced countries.

In technologically developed countries the three main types of vegetable farming are based on production of vegetables for the fresh market, for canning, freezing, dehydration, and pickling, and to obtain seeds for planting.

Production for the Fresh Market

This type of vegetable farming is normally divided into home gardening, market gardening, truck farming and vegetable forcing.

Home gardening provides vegetables exclusively for family use. About one-fourth of an acre (one-tenth of a hectare) of land is required to supply a family of six. The most suitable vegetables are those producing a large yield per unit of area. Bean, cabbage, carrot, leek, lettuce, onion, parsley, pea, pepper, radish, spinach, and tomato are desirable home garden crops.

Market gardening produces assorted vegetables for a local market. The development of good roads and of motor trucks has rapidly extended available markets; the market gardener, no longer forced to confine his operations to his local market, often is able to specialize in the production of a few, rather than an assortment, of vegetables; a transformation that provides the basis for a distinction between market and truck gardening in the mid-20th century. Truck gardens produce specific vegetables in relatively large quantities for distant markets.

In the method known as forcing, vegetables are produced out of their normal season of outdoor production under forcing structures that admit light and induce favourable environmental conditions for plant growth. Greenhouses, cold frames, and hotbeds are common structures used. Hydroponics, sometimes called soilless culture, allows the grower to practice automatic watering and fertilizing, thus reducing the cost of labour. To successfully compete with other fresh market producers, greenhouse vegetable growers must either produce crops when the outdoor supply is limited or produce quality products commanding premium prices.

Production for Processing

Processed vegetables include canned, frozen, dehydrated, and pickled products. The cost of production per unit area of land and per ton is usually less for processing crops than for the same crops grown for market because raw material appearance is not a major quality factor in processing. This difference allows lower land value, less hand labour, and lower handling cost. Although many kinds of vegetables can be processed, there are marked varietal differences within each species in adaptability to a given method.

Specifications for vegetables for canning and freezing usually include small size, high quality, and uniformity. For many kinds of vegetables, a series of varieties having different dates of maturity is required to ensure a constant supply of raw material, thus enabling the factory to operate with an even flow of input over a long period. Acceptable processed vegetables should have a taste, odour, and appearance comparable with the fresh product, retain nutritive values, and have good storage stability.

Vegetables Raised for Seed Production

This type of vegetable farming requires special skills and techniques. The crop is not ready for harvest when the edible portion of the plant reaches the stage of maturity; it must be carried through further stages of growth. Production under isolated conditions ensures the purity of seed yield. Special techniques are applied during the stage of flowering and seed development and also in harvesting and threshing the seeds.

Production Factors and Techniques

Profitable vegetable farming requires attention to all production operations, including insect, disease, and weed control and efficient marketing. The kind of vegetable grown is mainly determined by consumer demands, which can be defined in terms of variety, size, tenderness, flavour, freshness, and type of pack. Effective management involves the adoption of techniques resulting in a steady flow of the desired amount of produce over the whole of the natural growing season of the crop. Many vegetables can be grown throughout the year in some climates, although yield per acre for a given kind of vegetable varies according to the growing season and region where the crop is produced.

Climate

Climate involves the temperature, moisture, daylight, and wind conditions of a specific region. Climatic factors strongly affect all stages and processes of plant growth.

Temperature

Temperature requirements are based on the minimum, optimum, and maximum temperatures during both day and night throughout the period of plant growth. Requirements vary according to the type and variety of the specific crop. Based on their optimum temperature ranges, vegetables may be classed as cool-season or warm-season types. Cool-season vegetables thrive in areas where the mean daily temperature does not rise above 70 °F (21 °C). This group includes the artichoke, beet, broccoli, brussels sprouts, cabbage, carrot, cauliflower, celery, garlic, leek, lettuce, onion, parsley, pea, potato, radish, spinach, and turnip. Warm-season vegetables, requiring mean daily temperature of 70 °F or above, are intolerant of frost. These include the bean, cucumber, eggplant, lima bean, okra, muskmelon, pepper, squash, sweet corn (maize), sweet potato, tomato, and watermelon.

Premature seeding, or bolting, is an undesirable condition that is sometimes seen in fields of cabbage, celery, lettuce, onion, and spinach. The condition occurs when the plant goes into the seeding stage before the edible portion reaches a marketable size. Bolting is attributed to either extremely low or high temperature conditions in combination with inherited traits. Specific vegetable strains or varieties may exhibit significant differences in their tendency to bolt.

Young cabbage or onion plants of relatively large size may bolt upon exposure to low temperatures near 50° to 55 °F (10° to 13 °C). At high temperatures of 70° to 80 °F (21° to 27 °C) lettuce plants do not form heads and will show premature seeding. The fruit sets of tomatoes are adversely affected by relatively low and relatively high temperatures. Tomato breeders, however, have developed several new varieties, some setting fruits at a temperature as low as 40 °F (4 °C) and others at a temperature as high as 90 °F (32 °C).

Moisture

The amount and annual distribution of rainfall in a region, especially during certain periods of development, affects local crops. Irrigation may be required to compensate for insufficient rainfall.

For optimum growth and development, plants require soil that supplies water as well as nutrients dissolved in water. Root growth determines the extent of a plant's ability to absorb water and nutrients, and in dry soil root growth is greatly retarded. Extremely wet soil also retards root growth by restricting aeration. Atmospheric humidity, the moisture content of the air, also contributes moisture. Certain seacoast areas characterized by high humidity are considered especially adapted to the production of such crops as the artichoke and lima bean. High humidity, however, also creates conditions favourable for the development of certain plant diseases.

Daylight

Light is the source of energy for plants. The response of plants to light is dependent upon light intensity, quality, and daily duration, or photoperiod. The seasonal variation in day length affects the growth and flowering of certain vegetable crops. Continuation of vegetative growth, rather than early flower formation, is desirable in such crops as spinach and lettuce. When planted very late in the spring, these crops tend to produce flowers and seeds during the long days of summer before they attain sufficient vegetative growth to produce maximum yields. The minimum photoperiod required for formation of bulbs in garlic and onion plants differs among varieties, and local day length is a determining factor in the selection of varieties.

Each of the climatic factors affects plant growth, and can be a limiting factor in plant development. Unless each factor is of optimum quantity or quality, plants do not achieve maximum growth. In addition to the importance of individual climatic factors, the interrelationship of all environmental factors affects growth.

Certain combinations may exert specific effects. Lettuce usually forms a seedstalk during the long days of summer, but the appearance of flowers may be delayed, or even prevented, by relatively low temperature. An unfavourable temperature combined with unfavourable moisture conditions may cause the dropping of the buds, flowers, and small fruits of the pepper, reducing the crop yield. Desirable areas for muskmelon production are characterized by low humidity combined with high temperature. In the production of seeds of many kinds of vegetables, absence of rain, or relatively light rainfall, and low humidity during ripening, harvesting, and curing of the seeds are very important.

Site

The choice of a site involves such factors as soil and climatic region. In addition, with the continued trend toward specialization and mechanization, relatively large areas are required for commercial production, and adequate water supply and transportation facilities are essential. Topography—that is, the surface of the soil and its relation to other areas—influences efficiency of operation. In modern mechanized farming, large, relatively level fields allow for lower operating costs. Power equipment may be used to modify topography, but the cost of such land renovation may be prohibitive. The amount of slope influences the type of culture possible. Fields with a moderate slope should be contoured, a process that may involve added expense for the building of terraces and diversion ditches. The direction of a slope may influence the maturation time of a crop or may result in drought, winter injury, or wind damage. A level site is generally most desirable, although a slight slope may assist drainage. Exposed sites are not suitable for vegetable farming because of the risk of damage to plants by strong winds.

The soil stores mineral nutrients and water used by plants, as well as housing their roots. There are two general kinds of soils—mineral and the organic type called muck or peat. Mineral soils include sandy, loamy, and clayey types. Sandy and loamy soils are usually preferred for vegetable production. Soil reaction and degree of fertility can be determined by chemical analysis. The reaction of the soil determines to a great extent the availability of most plant nutrients. The degree of acid, alkaline, or neutral reaction of a soil is expressed as the pH, with a pH of 7 being neutral, points below 7 being acid, and those above 7 being alkaline. The optimum pH range for plant growth varies from one crop to another. A soil can be made more acid, or less alkaline, by applying an acid-producing chemical fertilizer such as ammonium sulfate.

The inherent fertility of soils affects production quantity, and a sound fertility program is required to maintain productivity. The ability of a soil to support plant life and produce abundant harvests is dependent on the immediately available nutrients in the soil and on the rate of release of additional nutrients that are present but not available to plants. The rate of release of these additional nutrients is affected by such factors as microbial action, soil temperature, soil moisture, and aeration. Depletion of soil fertility may occur as a result of crop removal, erosion, leaching, and volatilization, or evaporation, of nutrients.

Soil Preparation and Management

Soil preparation for vegetable growing involves many of the usual operations required for other crops. Good drainage is especially important for early vegetables because wet soil retards development. Sands are valuable in growing early vegetables because they are more readily drained than the heavier soils. Soil drainage accomplished by means of ditches or tiles is more desirable than the drainage obtained by planting crops on ridges because the former not only removes the excess water but also allows air to enter the soil. Air is essential to the growth of crop plants and to certain beneficial soil organisms making nutrients available to the plants.

When crops are grown in succession, soil rarely needs to be plowed more than once each year. Plowing incorporates sod, green-manure crops, and crop residues in the soil; destroys weeds and insects; and improves soil texture and aeration. Soils for vegetables should be fairly deep. A depth of six to eight inches (15 to 20 centimetres) is sufficient in most soils.

Soil management involves the exercise of human judgment in the application of available knowledge of crop production, soil conservation, and economics. Management should be directed toward producing the desired crops with a minimum of labour. Control of soil erosion, maintenance of soil organic matter, the adoption of crop rotation, and clean culture are considered important soil-management practices.

Soil erosion, caused by water and wind, is a problem in many vegetable-growing regions because the topsoil is usually the richest in fertility and organic matter. Soil erosion by water can be controlled by various methods. Terracing divides the land into separate drainage areas, with each area having its own waterway above the terrace. The terrace holds the water on the land, allowing it to soak into the soil and reducing or preventing gullying. In the contouring system, crops are planted in rows at the same level across the field. Cultivation proceeds along the rows rather than up and down the hill. Strip cropping consists of growing crops in narrow strips across a slope, usually on the contour. Soil erosion by wind can be controlled by the use of windbreaks of various kinds, by keeping the soil well supplied with humus, and by growing cover crops to hold the soil when the land is not occupied by other crops.

Maintenance of the organic-matter content of the soil is essential. Organic matter is a source of plant nutrients and is valuable for its effect on certain properties of the soil. Loss of organic matter is the result of the action of micro-organisms that gradually decompose it to carbon dioxide. The addition of manures and the growing of soil-improving crops are efficient means of supplying soil organic matter. Soil-improving crops are grown solely for the purpose of preparing the soil for the growth of succeeding crops. Green-manure crops, grown especially for soil improvement, are turned under while still green and usually are grown during the same season of the year as the vegetable crops. Cover crops, raised for both soil protection and improvement, are only grown during seasons when vegetable crops do not occupy the land. When a soil-improving crop is turned under, the various nutrients that have contributed to the growth of the crop are returned to the soil, adding a quantity of organic matter. Both legumes, those plants such as peas and beans having fruits and seeds formed in pods, and nonlegumes are effective soil-improving crops. The legumes, however, are more valuable, because they contribute nitrogen as well as humus. The rate of decomposition of plant material depends on the kind of crop, its stage of growth, and soil temperature and moisture. The more succulent the material is at the time it is turned under, the more quickly it decomposes. Because dry material decomposes more slowly than green material, it is desirable to turn under soil-improving crops before they are mature, unless considerable time is to elapse between the plowing and the planting of the succeeding crop. Plant material decomposes most rapidly when the soil is warm and well supplied with moisture. If soil is dry when a soil-improving crop is turned under, little or no decomposition will occur until rain or irrigation supplies the necessary moisture.

The chief benefits derived from crop rotation are the control of disease and insects and the better use of the resources of the soil. Rotation is a systematic arrangement for the growing of different crops in a more or less regular sequence on the same land. It differs from succession cropping in that rotation cropping covers a period of two, three, or more years, while in succession cropping two or more crops are grown on the same land in one year. In many regions vegetable crops are grown in rotation with other farm crops. Most vegetables grown as annual crops fit into a four-or five-year rotation plan. The system of intercropping, or companion cropping, involves the growing of two or more kinds of vegetables on the same land in the same growing season. One of the vegetables must be a small-growing and quick-maturing crop; the other must be larger and late maturing.

In the practice of clean culture, commonly followed in vegetable growing, the soil is kept free of all competing plants through frequent cultivation and the use of protective coverings, or mulches, and weed killers. In a clean vegetable field the possibility of attack by insects and disease-incitant organisms, for which plant weeds serve as hosts, is reduced.

Propagation

Propagation of crop plants, involving the formation and development of new individuals in the establishment of new plantings, is usually accomplished by the use of either seeds or the vegetative parts of plants. The first type, known as sexual propagation, is used for asparagus, bean, broccoli, cabbage, carrot, cauliflower, celery, cucumber, eggplant, leek, lettuce, lima bean, okra, onion, muskmelon, parsley, pea, pepper, pumpkin, radish, spinach, sweet corn (maize), squash, tomato, turnip, and watermelon. The second type, asexual propagation, is used for the artichoke, garlic, girasole, potato, rhubarb, and sweet potato.

Although seed cost is a small portion of the total cost of crop production, seed quality strongly affects crop success or failure. Good seed should be accurately labelled, clean, graded to size, viable, and free of diseases and insects. The reliability of the seed house is an important factor in obtaining good-quality seed. Viability, or ability to grow, and longevity, the period of viability, are characteristics of seeds of any vegetable kind. In cool, dry storage conditions, those vegetable seeds having comparatively short longevity of one to two years are okra, onion, parsley, and sweet corn. Seeds having three-year longevity are those of the asparagus, bean, carrot, leek, and pea; four-year longevity is characteristic of the beet, chard, pepper, pumpkin, and tomato seeds; longevity of five years characterizes the seeds of broccoli, cabbage, cauliflower, celery, cucumber, eggplant, lettuce, muskmelon, radish, spinach, squash, turnip, and watermelon. The dry seeds of all vegetables, when packed under vacuum in hermetically sealed cans, should remain viable for a longer period than seeds stored under less protective conditions.

Crops grown from hybrid seeds (the offspring of two or more selected parental varieties and known as F1) yield vegetables of high quantity and quality. The hybrid-seed industry is based on the production of new seed each year from the controlled pollination of selected parents found to produce the desired combination of characters in the progeny.

Planting

Most vegetable crops are planted in the field where they are to grow to maturity. A few kinds are commonly started in a seedbed, established in the greenhouse or in the open, and transplanted as seedlings. Asparagus seeds are planted in a seedbed to produce crowns used for field setting. Some vegetables can be either directly seeded in the field or grown from transplants. These include broccoli, cabbage, cauliflower, celery, eggplant, leek, lettuce, onion, pepper, and tomato. The time and method of planting seeds and plants of a particular vegetable influence the success or failure of the crop. Important factors include the depth of planting, the rate of planting, and the spacing both between rows and between plants within a row.

Factors to be considered in determining the time of planting include soil and weather conditions, kind of crop, and desired harvest time. When more than one planting of a crop is made, the second and later plantings should be timed to provide a continuous harvest for the period desired. The soil temperature required for germination of the planted seed varies markedly with the various kinds of vegetables. Vegetables that will not germinate at a temperature below 60 °F (16 °C) include the bean, cucumber, eggplant, lima bean, muskmelon, okra, pepper, pumpkin, squash, and watermelon. Temperatures higher than 90° F (32 °C) are not favourable for the germination of seeds of celery, lettuce, lima bean, parsley, pea, and spinach.

The quantity of seeds planted, or rate of planting, is mainly determined by the characteristics of the vegetable plant. The size of seeds affects the number of plants raised in a given area. Watermelon varieties, for example, differ in seed size expressed as weight. The Sugar Baby variety has an average weight of 1.4 ounces (41 grams) for 1,000 seeds; those of Blackstone variety average 4.4 ounces (125 grams). If the two are grown on two separate plots of the same area and 4.4 ounces of seeds of each cultivar are planted, the result would be three times as many of the Sugar Baby plants as the Blackstone type. Seed size and plant-growth pattern of a vegetable are major factors that govern the number of plants raised in a given area. The trend in the early 1980s was to increase plant population for many crops to achieve the greatest yield possible without impairing quality. As plant population increases per unit area, a point is reached at which each plant begins to compete for certain essential growth factors—e.g., nutrients, moisture, and light. When the population is below the level in which competition between plants occurs, increased population will have no effect on individual plant performance, and the yield per unit area will increase in direct proportion to the increment of population. When competition for essential growth factors occurs, however, yield per plant decreases.

Early harvest and economical use of space are the principal objectives of growing vegetable crops from transplants produced in a greenhouse or outdoor seedbed. It is easier to care for young plants of the cabbage, cauliflower, celery, onion, and tomato in small seedbeds than to sow the seeds in the place where the crop is to grow and mature. Land is free longer for another crop, and weeds, insects, diseases, and irrigation are more readily and economically controlled. The production of transplants is often a specialty of growers who sell their produce to other vegetable growers. The seeds may be planted at a rate three to six times that commonly used for a direct-seeded field. The young plants are removed for use as transplants when they reach the desired size and age, approximately 40 to 60 days after seeding.

Care of Crops during Growth

Practices required for a vegetable crop growing in the field include cultivation; irrigation; application of fertilizers; control of weeds, diseases, and insects; protection against frost; and the application of growth regulators if necessary.

Cultivation

Cultivation refers to stirring the soil between rows of vegetable plants. Because weed control is the most important function of cultivation, this work should be performed at the most favourable time for weed killing, when the weeds are breaking through the soil surface. When the plants are grown on ridges, it is necessary to cover the basal plant portion with soil in the case of such vegetables as asparagus, carrot, garlic, leek, onion, potato, sweet corn, and sweet potato.

Irrigation

Vegetable production requires irrigation in arid and semi-arid regions, and irrigation is frequently used as insurance against drought in more humid regions. In areas having intermittent rain for five or six months, with little or none during the remainder of the year, irrigation is essential throughout the dry season and may also be needed between rainfalls in the rainy season. The two types of land irrigation generally suited to vegetables are surface irrigation and sprinkler irrigation. A level site is required for surface irrigation, in which the water is conveyed directly over the field in open ditches at a slow, nonerosive velocity. Where water is scarce, pipelines may be used, eliminating losses caused by seepage and evaporation. The distribution of water is accomplished by various control structures, and the furrow method of surface irrigation is frequently employed because most vegetable crops are grown in rows. Sprinkler irrigation conveys water through pipes for distribution under pressure as simulated rain.

Irrigation requirements are determined by both soil and plant factors. Soil factors include texture, structure, water-holding capacity, fertility, salinity, aeration, drainage, and temperature. Plant factors include type of vegetable, density and depth of the root system, stage of growth, drought tolerance, and plant population.

Fertilizer Application

Soil fertility is the capacity of the soil to supply the nutrients necessary for good crop production, and fertilizing is the addition of nutrients to the soil. Chemical fertilizers may be used to supply the needed nitrogen, phosphorus, and potassium. Chemical tests of soil, plant, or both are used to determine fertilizer needs, and the rate of application is usually based on the fertility of the soil, the cropping system employed, the kind of vegetable to be grown, and the financial return that might be expected from the crop. Methods of fertilizer application include scattering and mixing with the soil before planting; application with a drill below the surface of the soil at the time of planting; row application before or at planting time; and row application during plant growth, also called side-dressing. Plowed down broadcast fertilizers have recently been used in combination with high analysis liquid fertilizers applied at planting or as a side-dressed band. Mechanical planting devices may employ fertilizer attachments to plant the fertilizer in the form of bands near the seed. For most vegetables, the bands are placed from two to three inches (five to 7.5 centimetres) from the seed, either at the same depth or slightly below the seed.

Weed Control

Weeds (plants growing where they are not wanted) reduce crop yield, increase production cost, and may harbour insects and diseases that attack crop plants. Methods employed to control weeds include hand weeding, mechanical cultivation, application of chemicals acting as herbicides, and a combination of mechanical and chemical means. Herbicides, selective chemical weed killers, are absorbed by the plant and induce a toxic reaction. The amount and type of herbicide that can be safely used to protect vegetable crops depends on the tolerance of the specific crops to the chemical. Most herbicides are applied as a spray, and the appropriate time for application is determined by the composition of the herbicide and the kind of vegetable crop to be treated. Preplanting treatments are applied before the crop is planted; preemergence treatments are applied after the crop is planted but before its seedlings emerge from the soil; and postemergence treatments are applied to the growing crop at a definite stage of growth.

Disease and Insect Control

The production of satisfactory crops requires rigorous disease and insect-control measures. Crop yield may be lowered by disease or insect attack, and when plants are attacked at an early stage of growth the entire crop may be lost. Reduction in the quality of vegetable crops may also be caused by diseases and insects. Grades and standards for market vegetables usually specify strict limits on the amount of disease and insect injury that may be present on vegetables in a designated grade. Vegetables remain vulnerable to insect and disease damage after harvesting, during the marketing and handling processes. When a particular plant pest is identified, the grower can select and apply appropriate control measures. Application of insect control at the times specific insects usually appear or when the first insects are noticed is usually most effective. Effective disease control usually requires preventive procedures.

Diseases are incited by such living organisms as bacteria, fungi, and viruses. Harmful material enters the plant, develops during an incubation period, and finally causes infection, the reaction of the plant to the pathogen, or disease-producing organism. Control is possible during the inoculation and incubation phases, but when the plant reaches the infection stage it is already damaged. Typical plant diseases include mildew, leaf spots, rust, and wilt. Chemical fungicides may be used to control disease, but the use of disease-resistant plant varieties is the most effective means of control.

Vegetable breeders have developed plant varieties resistant to one or more diseases; such varieties are available for the bean, cabbage, cucumber, lettuce, muskmelon, onion, pea, pepper, potato, spinach, tomato, and watermelon.

Insects are usually controlled by the use of chemical insecticides that kill through toxic action. Many insecticides are toxic to harmful insects but do not affect bees, which are valuable for their role in pollination.

Frost Protection

Frost protection may be accomplished by increasing the amount of heat radiated from the soil when frost is likely to occur. Irrigation on the day before a predicted frost provides additional moisture in the soil to increase the amount of heat given off as infrared rays. This extra heat protects the plants from frost injury. A continuous supply of water provided by sprinkler irrigation may also protect plants from frost. As the water freezes on the plant leaves, it loses heat that is absorbed by the plant leaves, maintaining leaf temperature at 32 °F (0 °C). Because of the sugars and other substances in plant cells, the freezing point of cell sap is somewhat lower than 32 °F.

Growth Regulators

It is sometimes desirable to retard or accelerate maturity in vegetable crops. A chemical compound may be applied to prevent sprouting in onion crops. It is applied in the field sufficiently early for absorption by the still-green foliage but late enough to avoid suppressing the bulb yield. Another substance may be used to end the dormancy, or rest period, of newly harvested potato tubers intended for planting. The treated seed potatoes have uniform sprout emergence. The same substance is applied to celery from two to three weeks before harvest to elongate the stalks and increase the yield and is also used to accelerate maturity in artichokes. A chemical compound, applied when adverse weather conditions prevail during the period of fruit setting, has been used to encourage fruit set.

Harvesting

The stage of development of vegetables when harvested affects the quality of the product reaching the consumer. In some vegetables, such as the bean and pea, optimum quality is reached well in advance of full maturity and then deteriorates, although yield continues to increase. Factors determining the harvest date include the genetic constitution of the vegetable variety, the planting date, and environmental conditions during the growing season. Successive harvest dates may be obtained either by planting varieties having different maturity dates or by changing the sequence of planting dates of one particular variety. The successive method is applicable to such crops as broccoli, cabbage, cauliflower, muskmelon, onion, pea, sweet corn (maize), tomato, and watermelon. Certain varieties of the carrot, celery, cucumber, lettuce, parsley, radish, spinach, or summer squash can be sown in succession throughout most of the year in some climates, thus prolonging the harvest period.

Hand harvesting is employed along with various mechanical aids for broccoli, cabbage, cauliflower, muskmelon, and pepper crops. Many vegetables grown for processing and some vegetables destined for the fresh market are mechanically harvested. Harvesting operations may be performed by a single machine in a single step for such vegetable crops as the bean, beet, carrot, lima bean, onion, pea, potato, radish, spinach, sweet corn, sweet potato, and tomato. Designers of harvesting machinery have been working to develop a multiple-picking harvester capable of adjustment for use with more than one crop. Vegetable breeders have been able to produce vegetables with characteristics suitable for machine harvesting, including compact plant growth, uniform development, and concentrated maturity.

Storage

Fresh vegetables are living organisms, and there is a continuation of life processes in the vegetable after harvest. Changes that occur in the harvested, nonprocessed vegetable include water loss, conversion of starches to sugars, conversion of sugars to starches, flavour changes, colour changes, toughening, vitamin gain or loss, sprouting, rooting, softening, and decay.

Some changes result in quality deterioration; others improve quality in those vegetables that complete ripening after harvest. Postharvest changes are influenced by such factors as kind of crop, air temperature and circulation, oxygen and carbon dioxide contents and relative humidity of the atmosphere, and disease-incitant organisms. To maintain the fresh vegetable in the living state, it is usually necessary to slow the life processes, though avoiding death of the tissues, which produces gross deterioration and drastic differences in flavour, texture, and appearance.

Storage of vegetables contributes to price stabilization by carrying over produce from periods of high production to periods of low production. It also extends the period of consumption of many kinds of vegetables. Storage conditions can contribute to the preservation of the natural living state of the edible portion and to the prevention of deterioration through control of temperature, relative humidity, and the quality of the produce to be stored. Vegetables for storage must be free from mechanical, insect, and disease injury and should be at the proper stage of maturity.

Common (unrefrigerated) storage and cold (refrigerated) storage are the methods generally employed for vegetables. Common storage, lacking precise control of temperature and humidity, includes the use of insulated storage houses, outdoor cellars, or mounds. Cold storage allows precise regulation of temperature and humidity and maintenance of constant conditions by use of a refrigeration and ventilation system. Temporary storage, suitable only for very brief storage periods, is frequently practiced in the shipping season when large lots are accumulated for carload or truck quantities. The refrigerator car or truck is a means of temporary storage used to protect produce while it is in transit. Short-term storage may last for four or six weeks. Economic factors, such as the probability that prices will increase later in the season, encourage long-term storage of such perishable vegetables as the onion, potato, and sweet potato.

Types of Vegetable Farming

The Different types of vegetable farming are:

1. Kitchen gardening/Home garden: It is the growing of vegetable crops in residential houses to meet the requirements of the family all the year around. Every individual is concerned with home or kitchen garden. Irrespective of the fact that the individual is a villager, a city dweller, live in town. Kitchen garden should be a future of his home.

2. Design of Kitchen Garden: Design of kitchen garden depends upon the character of the particular piece of land, its extent, situation etc. The following principles should be followed in designing the layout of kitchen garden:

- Location and site.

- Proper layout.

- Cropping pattern.

- Size 25 x 10 m for family of 5 persons.

- Shape should be rectangular and South east aspect is the most preferred for having more sun light.

3. For kitchen garden land should be selected in the backyard of the house (easier to work & make use of kitchen waste water.

- Layout of the garden should be such that it looks attractive and allow access to all the parts.

- The land should be laid out in small plots with narrow and path borders.

- In homes where no space is available one can grow vegetables in pots or boxes. Preference should be given to such vegetables which produce more number of fruits from an individual plant e.g. cucurbits, tomato, brinjal, chilli etc.

- Climbing type vegetables like cucurbits, pea beans etc. can be trained on the fences.

- Several sowings of one particular crop at short intervals should be done to ensure a steady supply of vegetables.

- Quick growing fruits trees like papaya, banana, lime etc. should be located on one side of the garden, preferably on Northern side so that there shading effect on the vegetables is on minimum side.

- Ridges which separate the beds should be utilized for growing root crops like radish, turnip, beet, carrot.

- Early maturing crops should be planted together in continuous row so that the areas may be available for putting next crop.

- The inter-space of some crops which are slow growing and take long duration to mature like cabbage, cauliflower, brinjal should be used for growing some quick growing crops like radish, turnip, palak, lettuce.

4. Market Gardening/Peri-urban vegetable farming: Peri-urban farming is known for its important role in providing self-employment besides enhancing the food security, helpful in poverty alleviation, waste management, community resource development and environmental sustainability:

- This is a type of garden which produces vegetables for local market.

- This type of garden was confined to the near vicinity of the cities when a quick transport was not developed.

- Most of the market gardens even today are located within 15-20 km of a city.

- The cropping pattern of these gardens will depend on the demand of the local market.

- The most important consideration is to develop a clearly focused marketing plan before any vegetable crops are planted.

- The land being costly, intensive methods of cultivation are followed.

- A market gardener will like to grow early varieties to catch the market early.

- He should be good salesman as he may have to sell his own produce.

- He must be a versatile person as he will have to grow a number of vegetables throughout the year.

- The high cost of land and labour is compensated by the availability of municipal compost, sludge and water near some cities and high return on the produce.

The preference of Indian consumers is mainly to have fresh and lush green vegetables and least for processed products. This provides a business opportunity to the growers living nearby the big cities or towns, generally referred as peri-urban areas to meet the requirement of consumers and earn higher profit. This production system focussing nearby big cities is also called as market gardening. Thus, peri-urban vegetable cultivation provides the possibility to cultivate a small piece of land on commercial line to generate income to meet the basic needs of a family.

Large quantity of solid waste is generated in cities during handling and marketing of fresh vegetable produce and otherwise also which in general creates health and environmental hazards. This can be recycled to produce manure for use in organic vegetable production.

Many farmers try to maximize their income by selling directly to consumers, thus bypassing wholesalers and other middlemen. Common marketing strategies can be adopted such as farmers stall in weekly vegetable market, roadside stands and sale agreement to restaurants, modern retail stores. Sometimes, organically grown vegetable produce in general get higher prices in the market. So, farmers may go for raising vegetable crops organically.

Considering the high cost and small size of farm land in the vicinity of a city and high cost of labour, water and energy, it is necessary for the farmer to have high productivity per unit area. Diversified crops are grown in peri-urban vegetable farms which also include specialty vegetables like red and yellow coloured sweet pepper, cherry tomato, broccoli, Brussels sprouts, baby corn, sweet corn, gherkin, leek, bunching onion, celery, parsley, chive, pak-choi, asparagus, artichoke etc. The specialty vegetables are becoming popular to meet the demands of consumers, restaurants and hotels in big cities.

Importance

- Efficient and effective use of land for growing essential vegetables for use of family.

- Saves some money as vegetables are quite costly in the market (fresh vegetables).

- Play important part in vegetable production.

- Constitute a very healthy hobby and the spare time of the family is well utilized.

- Kitchen gardening should be aimed at giving a continuous supply of vegetables to a family throughout the year.

- Pesticide residue free vegetables (health point of view).

- Training/education of children and to develop a sense of co-operation.

The other important considerations are choice of vegetables adapted to soil and climatic conditions, facilities of labour, water for irrigation and transport, proximity to market, and preferences of market and consumers. It is often profitable to have intercropping, succession of crops, relay cropping, mixed cropping and early maturing cultivars for continuous supply and for obtaining high price by bringing early produce in the market. Periurban production is either fast diminishing or moving farther from the city because of expansion of urban areas.

Truck Gardening

- This is a type of garden which produces special crops in relatively large quantities for distance markets.

- Truck gardens, in general, follow a more extensive and less intensive method of cultivation than market garden.

- The word truck has no relationship with a motor truck but it is derived from French word „troquer" means "to barter".

- The location of this type of garden is determined by the soil and climatic factors suitable for raising a particular crop.

- The commodities raised are usually sold through middle man.

- The truck gardener should be a specialized person.

- He should be proficient in large scale cultivation and production and handling of some special crops.

- He follows the mechanical method of cultivation hence cost of cultivation is less.

- The net income is also less as this includes the cost of transport and the charges of middle men.

- With the development of quick and easy transport system, the distinction between market and truck garden is continuously diminishing.

Vegetable Gardens for Processing

- These gardens come up around vegetable processing factories.

- Mainly responsible for regular supply of vegetables to factories.

- Earlier only a few factories existed which were dependent upon purchases from local markets.

- The end product from such local factories was not good from such a heterogeneous mixture.

- The prospects of future development are quite bright as people"s interest in the processing industry is growing.

- These gardens specialize in growing only a few vegetables in bulk.

- A heavier soil is chosen to obtain high and continuous yield rather than early yield.

- These gardens are required to grow particular varieties for canning, dehydration or freezing.

- The prices are paid on contract basis on weight and quantity of the produce.

- The return may be low but the cost of marketing and the transport charges are negligible.

Vegetable Forcing

In the method known as forcing, vegetables are produced out of their normal season of outdoor production under forcing structures that admit light and induce favourable environmental conditions for plant growth. Greenhouses, cold frames, and hotbeds are common structures used. Hydroponics, sometimes called soilless culture, allows the grower to practice automatic watering and fertilizing, thus reducing the cost of labour. To successfully compete with other fresh market producers, greenhouse vegetable growers must either produce crops when the outdoor supply is limited or produce quality products commanding premium prices.

- Tomato, cucumber and capsicum are commonly grown vegetables under these structures. These are mostly used during winter in the temperate regions. These crops cannot be grown without protection for their availability throughout year.

- River bed cultivation is a type of vegetable forcing i.e. growing of summer vegetables on river beds during winter months with the help of organic manures and wind breaks of dry grass.

- Sometimes, for early produce seedlings of tomato, brinjal, bell-pepper, chilli and cucurbits in poly-bags are forced to germinate in small protected structures.

Different Kinds of Vegetable Forcing

Protected Cultivation

It refers to agriculture with human interventions that create favourable conditions around the cultivated plants offsetting the detrimental effects of prevailing biotic and abiotic factors. Plants in open field conditions experience short cropping season, unfavourable climatic conditions (too cold, too hot, too dry and cloudy ambient) impairing photosynthetic activities, vulnerable to predators, pests, weeds, depleted soil moisture and plant nutrients. In protected agriculture one or more of these factors are controlled or altered, to the advantage of plants, where usually factors such as temperature, CO_2 concentration, relative humidity, access to insect and pest etc., are controlled to desirable limits. The factors controlled and range of control is decided by devises chosen and fitted on the structure. For economic reasons, protection or control is provided against the most significant stresses. Structures and environment control measurers employed separate this cultivated space and allowing cultivation in unfavourable ambient conditions in reasonably close to optimal conditions.

Advantages of Protected Cultivation

- Crop production with high productivity under unfavourable agro-climatic conditions.

- Productivity levels could be significantly higher (sometimes two-three times of that in open field agriculture).

- Quality of produce is usually superior because of isolation and controls.

- Higher input use efficiencies are achieved in the production of plant and animal products.

- Income per unit area significantly increases.

- Year-round production.

Production of crops under protected conditions has great potential in augmenting production and quality of vegetables, in main and also during off season and maximizing water and nutrient use efficiency under varied agro climatic conditions of the country. This technology has very good potential especially in peri- urban agriculture, since it can be profitably used for growing high value vegetable crops like, tomato, cherry tomato, coloured peppers, parthenocarpic cucumber, healthy and virus free seedlings production in agrientrepreneurial models.

Off Season Vegetable Cultivation under Walk-in-tunnels

Walk in tunnels are the temporary structures erected by using G.I. pipes and transparent plastic. Walk in tunnels are used for complete off season cultivation of vegetables like bottle gourd, summer squash, cucumber etc. during winter season (Dec.- mid February) the basic objective and utility of walk in tunnels is to fetch high price of the complete off season produce to earn more profit per unit area. The ideal size of a walk in tunnel of 4.0 m width and 30m length (l20 m²) and total cost of fabrication may be Rs.12000-14000/-.

Vegetable Gardens for Seed Production

- Good seed is the base of any successful farming industry.

- Seed production is a specialized field of vegetable growing.

- A thorough knowledge of the crop, its growth habit, mode of pollination, proper isolation distance are of prime importance for quality seed production.

- Specialized knowledge is required to handle the seed crop i.e. curing, threshing, cleaning, grading, packing and storage.

Types of Seeds

- Nucleus/breeder seed is produced by the person or organization which gives out the variety.

- Foundation seed is multiplied by government departments or by organization like NSC.

- Certified/Registered seeds usually multiplied by grower.

Floating Vegetable Gardens

- One more type of vegetable garden known as floating garden is seen on the Dal lake of Kashmir valley.

- Most of summer vegetables are supplied to Srinagar from these gardens.

- A floating base is made from the roots of typha grass which grow wild in some parts of lake.

- Once this floating base is ready, seedlings are transplanted on leaf compost made of vegetations growing wild in the lake.

- All the inter-cultural operations and occasional sprinkling of water are done from boats.

- This type of vegetable cultivation is a specialized technique and an art itself.

Organic Vegetable Gardening

In 1980, organic farming was defined by the USDA as a system that excludes the use of synthetic fertilizers, pesticides, and growth regulators.

Approaches and Production Inputs of Organic Farming

- Strict avoidance of synthetic fertilizers and synthetic pesticides.

- Crop rotations, crop residues, mulches.

- Animal manures and composts.

- Cover crops and green manures.

- Organic fertilizers and soil amendments.

- Biostimulants, humates, and seaweeds.

- Compost teas and herbal teas.

- Marine, animal, and plant by-products.

- Biorational, microbial, and botanical pesticides, and other natural pest control products.

- The Organic Foods Production Act, a section of the 1990 Farm Bill, enabled the USDA to develop a national program of universal standards, certification accreditation, and food labeling.

- In April 2001, the USDA released the Final Rule of the National Organic Program. This federal law stipulates, in considerable detail, exactly what a grower can and cannot do to produce and market a product as organic.

Container Gardening

In urban areas mainly in big cities, land is a big constraint for home/kitchen garden, many types of vegetables can be grown well in containers and space available in backyard, terrace, varandah, balcony can be utilized for this purpose where sunshine is easily available. Start with large enough pots. The 14 inch pots are plenty large for brinjal and cucumber and the 20-inch pots worked out well for tomatoes. Generally we should grow those vegetables which facilitate multiple harvests like tomato, leafy vegetables etc. instead of single harvest like cabbage or cauliflower etc.

Vegetable Diseases

1. In vegetable crops, Damping off, root rot, leaf spots, powdery mildew and downy mildew and virus diseases (Mosaic and leaf curl) cause severe yield loss.

Damping off: Pythium aphanidermatum is observed in most of the vegetables (Chillies, Tomato, Brinjal, Cabbage, Onion).

Symptoms: Occurs as pre emergence and post emergence damping off. In the pre-emergence the phase the seedlings are killed before emergence where young radical and the plumule are killed leads to rotting of the seedlings. The post-emergence phase is characterized by the infection of the young seedlings which become soft and water soaked at the collar region at the ground level and leads to toppling or collapse of the seedlings.

Favourable conditions: pathogen Survives in soil and excess soil moisture favours disease.

Management:

- Provide raised seed bed in nursery and fumigate with formaldehyde.

- Treat the seeds with Thiram @ 3 g/kg seed or Trichoderma viride @ 4 g kg seed or Pseudomonas fluorescens @ 10 g /kg of seed 24 hours before sowing.

- Soil application of Pseudomonas fluorescens @ 2.5 kg/ha mixed with 50 kg of FYM.

- Avoid stagnation of water.

- Drench with Copper oxychloride @ 2.5 g/l or Bordeaux mixture 1%.

2. Wilt: Wilt caused by Fusarium sp affects most of the vegetable crops:

- Tomato -Fusarium oxysporum f. sp. lycopersici.

- Chillies-Fusarium oxysporum f. sp. vesicatoria.

- Water melon- Fusarium oxysporum f. sp. Niveum.

Symptoms: Wilting of the plant is characterised by an initial of yellowing of the upper leaves that turn yellow and droop. Finally the vascular system of the plant is discoloured, particularly in the lower stem and roots and plants ecome wilted.

Favourable conditions: Soil and seed borne. Survives in soil for more than 10 years.Spreads through irrigation, farm implements.

Management:

- Drenching with 1% Bordeaux mixture or Copper oxy chloride @ 2.5 g/lit.

- Seed treatment with Pseudomonas fluorescens Pf1@ 10 g /kg of seed, followed by nursery application of Pf1@ 20 g/m² and seedling dip with Pf1 @ 5g/ l along with soil application of Pf1 @ 2.5 kg mixed with 50 kg FYM /ha at 30 days after transplanting.

- Spot drench with Carbendazim @ 1g/lit for wilt affected plants.

3. Leaf spots: Leaf spots caused by Cercospora sp., Septoria sp., Alternaria sp. and Botrytis sp occurs at all stages of the crop and leads to yield loss.

- Chillies: Cercospora capsisci.

- Tomato: Septoria lycopersici.

- Brinjal: Cercospora solani –melongenae, C.solani, Alternaria melongenae, A. Solani.

- Onion leaf blight and Purple blotch: Botrytis sp., Alternaria porri.

Symptoms: Leaf spots are brown and circular with small to large light grey centres and dark brown margins or dark black irregular spots with concentric rings or specks which coalese and cause drying and defoliation. Stem, petiole and pod lesions also have light grey centres with dark borders. Severely infected leaves drop off prematurely resulting in reduced yield. Favorable conditions: High humidity, drizzling, wind favors disease development and spread. Survives in infected crop debris.

Management:

- Seed treatment with Carbendazim @2g/kg of seed.

- Spraying Mancozeb @ 2 g/lit or Copper oxychloride @ 2.5 g/lit.

- For purple blotch of onion,spray Tebuconazole @ 1.5 ml/lit. or Mancozeb @ 2g/lit. or Zineb @ 2g/lit.

Virus Diseases - Mosaic and Leaf Curl

Mosaic: Transmitted by aphids. Affected plant appears stunted with light and dark green mottling on the leaves that becomes distorted, puckered and smaller than normal leaves. This disease is common in Tomato, Chillies, Brinjal, Gourds.

Leaf curl: Transmitted by white fly. Leaves are curled have margins that curl upward, giving them a cup-like appearance, reduced in size, yellowing between the veins, shortened internodes giving the plant a stunted appearance. This disease is common in Tomato and Chillies.

Management:

- Protected nursery in net house or green house.

- Placing yellow sticky traps @ 12/ha to monitor the vectors Raising barrier crops, cereals around the field.

- Raise two rows of Maize or Sorgnum for every 5 rows of chilli.

- Removal of weed host regularly Spray Imidacloprid @ 0.5 ml/lit. or Dimethoate @ 0.5 ml/lit. or Monocrotophos @ 1.5 ml/lit. or Acephate @ 1g/lit. at 15, 25, 45 days after transplanting to control vector.

- Virus perpetuates in cucurbits, legumes, pepper, tobacco, tomato and weed hosts so care should be taken.

Diseases of Tomato

1. Early blight: Alternaria solani Symptoms - Small, black spots enlarge and concentric rings in a bull's eye pattern can be seen in the center of the diseased area. Tissue surrounding the spots may turn yellow. Lesions on the stems may occur and girdle. On fruis, dark brown concentric rings are seen that affects the market quality. Shedding of immature fruits occur.

Fvaourable conditions: Survives in seeds and soil. High soil moisture creates high humidity that favors disease development initially on lower leaves.

Management:

- Removal of infected plant debris. Use of disease free seeds . Crop rotation with non solanaceoues crops. Spraying Mancozeb @ 2g/lit. or copper oxychloride @ 2.5 g/lit twice at 15 days interval.

2. Late blight: Phytophthora infestans.

Symptoms: Leaves stem and fruits are attacked Lesions appear as purplish to brown colour whicg leads to blighting under humid conditions. Marbbled areas on green fruits which later becomes brown and completely shriveled.

Favorable conditions: Pathogen survives in infected debris. Disease occurs in rainfed crops unde irrigation where dew is frequent and develops quickly in ariny season accompanied with high humidity.

Management:

- Crop rotation.

- Over head irrigation to bre avoided. Sparyinmg with mancozeb 0.2% or captafol @ 0.2% or Metalaxyl 0.2% or copper oxychloride @ 0.2%.

3. Bacterial wilt: Burkholderia solancearum.

Symptoms: Young seedling show yellowing and wilt. Curling of leaves occurs. More adventitious roots are formed. Later black discoloration of vascular tissues with gummy bacterial ooze is the characteristic symptom of severely infected plant.

Favorable conditions: Survives in soil and in infected plant debris spreads through irrigation water and farm implements.

Management:

- Crop rotation. Spraying with Agrimycin-100 @ 0.1g /lit. thrice at 10 days intervals effectively controls the disease.

4. Peanut bud necrosis disease: Groundnut bud necrosis virus.

- Numerous small, dark, circular spots appear on younger leaves, bronzing of leaves that later turn dark brown and wither. Plants affected at early stages do not flower, those afected at later stages show reuced flowers and poor fruit set and the growing shoots shows necrosis and finally death of plants.

- Fruits show numerous spots with concentric, circular markings.

- On ripe fruit, these markings alternate with bands of red and yellow.

- The spotted wilt virus is transmitted by thrips.

Management:

- Selection of healthy seedlings and rouging of infected plants up to 45 days of planting.

- Apply Carbofuran 3 G @1 kg a.i./ha in nursery at sowing and second application at 1.25 kg a.i./ha 10 days after transplanting in mainfield and 3 sprays of Dimethoate @1 ml/l or Methyl demeton @1 ml/l or Phosphomidan @ 1.0 ml/l at 25, 40 and 55 days after transplanting.

- Spraying Pseudomonas fluorescens @ 5g/l at 25th and 45th day after planting.

- The affected plants should be periodically removed and destroyed.

- Alternate or collateral hosts harboring the virus have to be removed.

- Raise barrier crops – Sorghum, Maize, Bajra at 5-6 rows around the field one month before planting.

Diseases of Chillies

1. Fruit Rot and Die Back: Colletotrichum capsici.

Symptoms: Necrosis of tender twigs from the tip backwards, profuse shedding of flowers. Drying up spreads from the flower stalks to the stem and subsequently causes die-back of the branches and stem which later on wither. Partially affected plants bear few fruits of low quality. Fruits rot appears on ripe fruits where small circular spost initially appear and spreads as oblong black greenish colour or markedly delimited by a black or straw colored area. Badly diseased fruits turn straw coloured with less pungency. The diseased fruist later shrivels nad dries up.

Favorable conditions: Seed borne and secondary spread by wind during periods of high humidity accompanied with rain. Disease appears after rains has stopped nad when there is prolonged deposition of dew on plant.

Management:

- Use disease free seeds Seed.

- treatment with Thiram or Captan @ 4g/kg.

- Spray Mancozeb @ 2 g/lit or Copper oxychloride @ 2.5 g/lit thrice at 15 days interval starting from noticing the die-back symptoms or at 60 days after planting.

2. Powdery mildew: Leveillula taurica.

Symptoms: White powdery growth on lower side of leaves leads to shedding of foliage causing severe reduction in fruit yield.

Management:

- Spray Wettable sulphur @ 3 g/lit or Carbendazim @ 1 g/lit, 3 sprays at 15 days interval from the first appearance of symptom

3. Bacterial leaf spot: Xanthomonas campestris pv. Vesicatoria Symptoms. The leaves exhibit small circular or irregular, dark brown or black greasy spots that form irregular lesions. Severely affected leaves become chlorotic and fall off.Stem infection leads to formation of cankerous growth and wilting of branches.On the fruits round, raised water soaked spots with a depression in the centre where in shining droplets of bacterial ooze is observed.

Management:

- Field sanitation and crop rotation Spray seedlings with Bordeaux mixture @ 1% or copper oxychloride @ 2.5g.

Diseases of Brinjal

Bacterial Wilt: Pseudomonas Solanacearum

Symptoms: Lower leaves may droop, yellowing of the foliage, stunting, wilting and finally collapse of the entire plant. The vascular system becomes brown and bacterial ooze comes out from the affected parts.

Favourable conditions: Presence of root knot nematode Meloidogyne incognita increases the disease incidence. Survives in soil and in infected plant debris. Spreads through irrigation water and farm implements.

Management:

- Crop rotation with cruciferous vegetables such as cauliflower.

- Fields should be kept clean and affected parts are to be collected and burnt.

- Spray 2% Bordeaux mixture or Spraying with Agrimycin-100 @ 100 ppm (0.1g/litre) thrice at 10 days intervals effectively controls the disease.

Phomopsis Fruit Rot: Phomopsis Vexans

Symptoms: Initially seen as blight on young seedlings. And girdle the plant as a result plant topples. On mature plant, stem lesion appear as dark brown and dpeads to entire stem, colaeses nad spreads to fruits. Fruits show soft watery lesion and later become mummified with minute black structure called pycnidia and remains dried.

Favorable conditions: Survives in soil in infected plant debris. Spreads though irrigation water and farm implements.

Management:

- Deep summer ploughing, crop rotation.

- Spraying Carbendazim + Mancozeb @ 0.1% or copper oxychloride @2 0.2% or difolotan 0.2%

Little leaf of Brinjal: Caused by Phytoplasma. Transmitted by Leaf hopper Hishimonas phycitis.

Symptoms: Reduction in leaf size. As disease progresses leaves become smaller, croded and the petioles become shorter. Later leaves become thin narrow and glabrous. Plant gives bushy appearance. Floral parts are modified in to green structuresand fail to produce fruit, If at all fruits are produced, fruits never mature and remain mummified.

Management:

- Removal of weed hosts.

- Spraying 100 ppm of Tetacycline or copper hydroxide @ 500 ppm.

Diseases of Gourds

Downy Mildew: Pseudoperonospora Cubensis

Symptoms: Disease affects pumpkin, Snake gourd, Ribbed gourd, Bottle gourd, Bitter gourd and Ash gourd. Whitish growth of fungus is seen on lower surface of leaves and corresponding upper leaves show pale green areas separated by dark green areas. The entire leaf dries up quickly.

Favorable conditions: High humidity with drizzling, low temperature of 15 to 25 °C accompanied with drizzling and dew favours disease.

Management:

- Seed treatment with Metalaxyl @ 2 g/kg. seeds.

- Spraying with Mancozeb @ 2g/lit. or Chlorothalonil @ 2 ml/lit. or Metalaxyl + Mancozeb @ 1g/lit.

Powdery Mildew: Erysiphe Cichoracearum

Symptoms: Disease occurs in Pumpkin, Snake gourd, Ribbedgourd, Bottle gourd, Bitter gourd and Ash gourd. White or brown growth on upper and lower surfaces of leaves and stems and leads to drying.

Favorable conditions: Dry season and high temperature. Favours disease development. Survives in infected plant debris.

Management:

- Spray Dinocap @ 1 ml/lit. or Carbendazim @ 0.5 g/lit.

- Mosaic : Cucumber mosaic virus.

Symptoms: Infected plants show cupping of leaves downward, severe mottling with alternating light green and dark green patches. Plants are stunted, and fruits are covered with bumpy protrusions. Severely affected cucumber fruit may be almost entirely white. The virus is readily transferred by aphids and survives on a wide variety of plants.

Management:

- Removal of weed host.

- Spray Dimethoate @ 2ml/lit. or Monocrotophos @ 1.5 ml/lit. or Acephate @ 1g/lit. to control insect vectors.

- Place yellow sticky traps @ 12/ ha.

Diseases of Onion

Basal Rot: Fusarium oxysporum f.sp. cepae.

Symptoms: Leaves turn yellow and then drying of leaf tip downwards. The bulb of the affected plant shows soft rotting and the roots get rotted. Whitish mouldy growth on the scales may be seen. This disease begins in the field and continues in storage.

Favorable conditions: Survives in soil in infected plant debris. Spreads though irrigation water and farm. Stagnation of water and temperature of 28 to 32 °C are favourable for disease development.

Management:

- Treat bulbs with Trichoderma viride @ 4g/kg. and apply T. viride 2 2.5 kg/ha. basally along with VAM 12.5 kg/ha.

- Soil drenching with Copper oxychloride @2.5g.lit.

Purple Blotch of Onion: Alternaria Porri

Symptoms: Occurs mainly on top of leaves. Infection starts with minute white dots with irregular chlorotic spots on the tip of leaves which later turn circular to oblong concentric balck velvety in the chlorotic area. Later spots coalesce and dry from tip downwards and hang down. Infection is laso seen on outer scales of bulb. Premature drying of foliage results in poor development of bulbs.

Favorable conditions: Carried through seed bulbs collected from infected area. Spreads through air borne spores. Warm humid weather, rain and dew favors disease development.

Downy Mildew: Peronospora Destructor

Symptoms: White downy growth appears on the lower surface of the leaves with chlorotic areas on upper surface.and finally the infected leaves are dried. The flowers are affected, dried and drop off. Entire plant is not killed only under sized bulbs are formed. Infected bulbs reamin small and succulent. Fungus invades flaoral parts and hence a small portion of seeds are also affected.

Favorable conditions: Pathogen requires cool, moist nights and moderate warm temperature with cloudy day for disease development. Survives in infected bulbs and in soil as oospores.

Management:

- Three sprays with Mancozeb @ 2 g/lit. Spraying should be started 20 days after transplanting and repeated at 10-12 days interval.

References

- Types-of-vegetables: cropsreview.com, Retrieved 13 August, 2019

- Cultivation, potato: fao.org, Retrieved 4 February, 2019

- Staub, Jack E. (2010). Alluring Lettuces: And Other Seductive Vegetables for Your Garden. Gibbs Smith. p. 230. ISBN 978-1-4236-0829-5

- Tomato, crop-information, land-water: fao.org, Retrieved 24 May, 2019

- Krech III, Shepard; Merchant, Carolyn; McNeill, John Robert, eds. (2004). Encyclopedia of World Environmental History. 3: O–Z, Index. Routledge. pp. 1071. ISBN 978-0-415-93735-1

- Onion, crop: indiaagronet.com, Retrieved 20 January, 2019

- Vegetable-farming: britannica.com, Retrieved 11 July, 2019

- Riotte, L. (1998). Carrots Love Tomatoes: Secrets of Companion Planting for Successful Gardening. Storey Publishing, LLC. p. 10. ISBN 978-1-60342-396-0

Chapter 3

Understanding Pomology

The branch of botany that deals with the study and cultivation of fruits is termed as pomology. It is concerned with the enhancement, cultivation, development and physiological studies of fruit trees. It aims to enhance the fruit quality, reduce the cost of production and regulate the production period. This chapter has been carefully written to provide an easy understanding of pomology and the different fruit crops which are studied under it.

Pomology is a branch of horticulture which focuses on the cultivation, production, harvest, and storage of fruit, especially tree fruits. Fruit orchards can be found all over the world, and tree fruits are a major industry in many countries, making pomology especially vital.

Any number of fruit trees can be included in a survey of pomology, like apricots, pears, plums, peaches, cherries, nectarines, and avocados. Pomologists also research tree nuts like almonds, walnuts, and pecans, among others.

One of the most critical aspects of pomology is the development of new fruit cultivars. A pomologist can cross-breed various fruit cultivars for specifically desired traits, such as flavor, hardiness, or disease-resistance. Pomology has contributed a number of exotic and interesting fruit cultivars to the world, such as the pluot, a cross between a plum and an apricot. If a pomologist can breed a distinctive and entirely new cultivar, he or she stands to profit significantly from the resulting patents.

Pomologists also look at the best way to grow trees, determining which regions trees grow in, and the amounts of water and fertilizer preferred by different cultivars. In addition, they study pests which attack fruit trees, and address issues of regional concern, like droughts or seasonal flooding.

Once a tree fruits, the work of a pomologist isn't over. Pomology is also used to develop new ways to harvest, store, and ship fruit, with a focus on keeping fruit healthy and flavorful until it reaches the consumer. Many cultivars have been specifically bred for easier harvesting and storage, but pomologists also work on things like agricultural equipment and special shipping containers for delicate fruits such as peaches.

Fruit Crops

Fruit Crop is any one of a group of wild and cultivated trees, shrubs, subshrubs, perennial undershrubs, and lianas that yield fleshy or hard edible fruits. In the USSR, plants from more than 50 genera, embracing 26 families, are used as fruit crops. The most important are the apple, pear, quince, sorbus, loquat, cherry plum, sweet cherry, plum, apricot, almond, peach, strawberry, and raspberry. Also important are the cornelian cherry, oleaster, sea buckthorn, European walnut, pecan, hazelnut, pistachio, olive, persimmon, fig, and mulberry. The pomegranate, feijoa, tangerine, orange, lemon, grapefruit, and citron are valuable for their fruit. Also important are the currant, gooseberry, European hazel, Actinidia, Schizandra, honeysuckle, viburnum, barberry, avocado and date palm. Many subtropical and tropical fruit crops are cultivated abroad. These include the mango, breadfruit tree, papaya, and banana.

Fruit crops may be deciduous (for example, berry, nut, pip [except for Eriobotrya], and drupe [except for cherry laurel] plants; some subtropical species, such as fig and persimmon) or evergreen (for example, raspberry, feijoa, all citrus). Fruit crops of temperate climates (apple, pear, cherry plum, plum) have a long winter dormancy. The characteristics of deciduous and evergreen crops determine the methods of cultivation to be used.

In the USSR, up to 90 percent of all orchard land is occupied by apple, cherry plum, plum, apricot, and pear trees. Sweet cherries, peaches, and quinces are cultivated in lesser quantities (2–3 percent). Nut crops (European walnut, hazelnut, pistachio, European hazel, and almond) occupy about 4 percent of orchard land, berry crops (strawberry, raspberry, currant, and gooseberry) about 2–3 percent, and subtropical fruits (olive, persimmon, fig, and pomegranate) and citrus (tangerine, orange, and lemon) about 1 percent.

The principal fruit crops in the USSR are the apple, cherry plum, and plum. The apple is the main fruit crop of Europe, Canada, the United States, and Argentina. In Mediterranean countries the olive, citrus fruits, and nuts are most widely cultivated. Subtropical and tropical fruits predominate in India and China, and bananas are the major fruit crops of Africa and South America. Fruit crops require various types of soil and climate.

In the northern and central fruit-growing zones of the USSR, apples, cherry plums, plums, pears, and all kinds of berries are cultivated. In the southern zone, in addition to the above-mentioned crops, the quince, sweet cherry, peach, and apricot are raised. Also cultivated are all kinds of nut crops. In the subtropical zone the principal fruit crops are the olive, fig, pomegranate, persimmon, and citrus.

Fruit crops are propagated mainly by vegetative methods. Such methods include grafting onto a stock, rooting cuttings, (currant, gooseberry, pomegranate, fig, olive), and shooting (cherry plum, plum).

Fruits can be classified on a botanical basis and for several operational purposes:

Botanical Classification

Fruits exhibit a variety of apparent differences that may be used for classification. Some fruits are borne on herbaceous plants and others on woody plants. A very common operational way of classifying fruits is according to fruit succulence and texture on maturity and ripening. On this basis there are two basic kinds of fruits-fleshy fruits and dry fruits. However, anatomically, fruits are distinguished by the arrangement of the carpel from which they developed. A carpel is sometimes called the pistil (consisting of a stigma, style, and ovary), the female reproductive structure.

A fruit is mature ovary. The ovary may have one or more carpel. Even though the fruit is a mature ovary, some fruits include other parts of the flower and are called accessory fruits. Combining carpel number, succulence characteristics, and anatomical features, fruits may be classified into three kinds, simple, multiple, or aggregate.

Simple fruits develop from a single carpel or sometimes from the fusing together of several carpel. This group of fruits is very diverse. When mature and ripe, the fruit may be soft and fleshy, dry and woody, or have a papery texture. There are three types of fleshy fruits.

Fleshy Fruits

1. Drupe: A drupe may comprise one to, several carpels. Usually, each carpel contains one seed. The endocarp (inner layer) of the fruit is hard and stony and is usually highly attached to the seed. Examples are cherry (Prunusspp.), olive, coconut (Cocos nucifera), peach (Prunus persica), and plum (Prunus domestica).

2. Berry: A berry is a fruit characterized by an inner pulp that contains a few to several seeds but not pits. It is formed from one or several carpels. Examples are tomato (Lycopersiconesculentum), grape (Vitis spp.), and pepper (Capsicum anuum). If the exocarp (skin) is leathery and contains oils, as in the citrus fruits (e.g., orange (Citrus sinensis), lemon (Citrus lemon), and grapefruit [Citrus paradisi]), the berry is called a hesperidium. Some berries have a rind, as in watermelon (Citrullus vulgaris), cucumber (Cucumis sativus), and pumpkin (Cucurbita pepo). This type of a berry is called a pepo.

3. Pome: A pome is a pitted fruit with a stony interior. The pit usually contains one seed chamber and one seed. This very specialized fruit type develops from the ovary, with most of the fleshy part formed from the receptacle tissue (the enlarged base of the perianth). Pomes are characteristic of one subfamily of the family Rosaceae (rose family). Examples of pomes are apple (Pyrus malus), pear (Pyrus communis), and quince (Cydonia oblonga).

4. Dry fruits: Dry fruits are not juicy or succulent when mature and ripe. When dry, they may split open and discharge their seeds (called dehiscent fruits) or retain their seeds (calledindehiscent fruits).

5. Dehiscent Fruits: A fruit developed from a single carpel may split from only one side at maturity to discharge its seeds. Such a fruit is called a follicle. Examples are columbine (Aquilegia spp.), milkweed (Asclepias spp.), larkspur (Delphinium spp.), and magnolia (Magnolia spp.). Sometimes, the splitting of the ovary occurs along two seams, with seeds borne on only one of the halves of the spilt ovary. Such a fruit is called a legume, example being pea (Pisum sativum), bean (Phaseolus vulgaris), and peanut (Arachis hypogaea). In a third type of dehiscent fruit, called silique or silicle, seeds are attached to central structure, as occurs in radish (Raphanus sativus) and mustard (Brassica campestris). The most common dehiscent simple seeds are discharged when the capsule splits longitudinally. In some species, seeds are discharged when the capsule splits longitudinally. In others, seeds exit through holes near the top of the capsule, such as in lily (Lilum spp.), iris (Iris spp.), and poppy (Papaverspp.).

6. Indehiscent Fruits: Some indehiscent fruits may have a hard pericarp (exocarp + mesocarp + endocarp). This stony fruit wall is cracked in order to reach the seed. Such fruits are called nuts, as found in chestnut (Castanea spp.) and hazelnut (Corylus spp.). Nuts develop from a compound ovary. Sometimes the pericarp of the fruit is thin and the ovaries occur in pairs, as found in dill (Anethum graveolens) and carrot (Daucus carota). This fruit type is

called a schizocarp. In maple (Acer spp.), ash (Fraxinus spp.), elm (Ulmus spp.), and other species, the pericarp has a wing and is called a samara. Where the pericarp is not winged but the single seed is attached to the pericarp only at its base, the fruit is called an achene. Achenes are the most common indehiscent fruits. Examples are the buttercup family (Ranuculaceae) and sunflower. In cereal grains (Poaceae or grass family), the seed, unlike in an achene, is fully fused to the pericarp. This fruit type is called a caryopsis or grain.

Other Operational Classifications

Fruits may also be classified according to other operational uses:

1. Temperate fruits or tropical fruits: Temperature fruits are fruits from plants adapted to cool climates, and tropical fruits are produced on plants adapted to warm climates. For example, apple (Pyrus malus), peach {Prunus domestica) are temperate fruits, whereas mango (Mangifera indica) and coconut (Cocos nucifera) are tropical fruits.

2. Fruit trees: Tree fruits are fruits borne on trees, such as apple (Pyrus malus) and mango (Mangifera indica).

3. Small fruits: Small fruits are predominantly woody, perennial, dicot angiosperms. They are usually vegetatively propagated and bear small- to moderate-sized fruits on herbs, vines, or shrubs. Examples are grape (Vitis spp.), strawberry (Fragaria spp.), and blackberry (Rubus spp.). Small fruits require training and pruning (removal of parts of the shoot) to control growth and remove old canes (branches) to obtain desired plant shape and high productivity.

4. Bramble fruits: Bramble fruits are non-tree fruits that usually require physical support (such as a trellis) during cultivation. Examples are raspberry (Rubus spp.), blackberry (Rubus spp.), and boysenberry) (Rubus spp.). Bramble fruits also require training and pruning in cultivation.

Scope and importance of fruit crops Fruit growing is one of the important and age old practices. Cultivation of fruit crops plays an important role in overall status of the mankind and the nation. The standard of living of the people of a country is depending upon the production and per capita consumption of fruits. Fruit growing have more economic advantages.

Economic Importance

1. High productivity: High yield per unit area: From a unit area of land more yield is realized from fruit crops than any of the agronomic crops. The average yields of Papaya, Banana and Grapes are 10 to 15 times than that of agronomic crops.

2. High net profit: Through, the initial cost of establishment of an orchard is high, it is compensated by higher net profit due to higher productivity or high value of produce.

    Eg- Wheat/GN/Ragi- 3.0 -4.0 tonnes/ha-25-35,000-00,

    Grapes/Mango/Banana-20-40t/ha-1.5-2.5 lakh/ha.

3. Source of raw material for agro based industries: Fruit farming provides raw materials for various agro based industries- canning and preservation (fresh fruits), coir industries (coconut husk), pharmaceutical industry (Aonla, Papaya, Jamun) Transporting and packaging industries etc.

4. Efficient utilization of resources: Growing of fruits being perennial in nature, enables grower to remain engaged throughout the year in farm operations and to utilize fully the resources & assets like machinery, labour, land water for production purpose throughout the year compared to agronomic crops.

5. Utilization of waste and barren lands for production: Although, most of the fruits crops require perennial irrigation and good soil for production, there are many fruit crops of hardy in nature, Mango, Ber, Cashew, Custard apple, Aonla, Phalsa, Jamun etc. which are grown on poor shallow, undulated soils considered unsuitable for growing grain/agronomical crops.

6. Foreign exchange: Many fresh fruits, processed products and spices are exported to several countries earning good amount of foreign exchange.

Nutritional Importance

1. Importance of fruits in human diet is well recognized. Man cannot live on cereals alone.

2. Fruits and vegetables are essential for balanced diet and good health.

3. Nutritionist advocates 60-85g of fruits and 360 gm.

4. Vegetables per capita per day in addition to cereals, pulses, egg etc.

5. Fruits and vegetables are good sources of vitamins and minerals without which human body cannot maintain proper health and develop resistance to disease they also contain pectin, cellulose, fats, proteins etc.

Mango

Mango (Mangifera Indica) is the most ancient among the tropical fruits and believed to have originated in the Indo – Burma region. India is the major mango producing country in the world with an annual production of 8.50 million tonnes from an area of one million hectares. Mango is basically a tropical plant but endures wide range of temperature. It grows well under tropical and sub-tropical conditions. It gives profitable yield in semi-arid conditions, especially with irrigation.

Mango Varieties In according to the climatic conditions, lots of mangoes type are present in the world. All these were prepared for high-yielding mango fruit production.

Below is a list top 10 hybrid mango varieties which are the sweetest mango in the world:

1. ALPHONSO (Gujarat, Maharashtra & other Western parts of India).

2. VALENCIA PRIDE (South Florida).

3. BADAMI (India).

4. CHAUNSA (Punjab, Pakistan).

5. NAM DOK MAI or GOLDEN MANGO (Thailand).

6. GLENN (Florida).

7. SINDHRI (Pakistan, Sindh area).

8. MADAME FRANCIQUE or DESERT MANGO (Haiti).

9. KESAR (Gujarat & other Indian States).

10. KEITT (Found in Miami).

Climate Condition for Mango Farming

As mango trees are vigorous in nature & require less maintenance than other commercial fruit tree farming business. So, mango cultivation can be done anywhere, where there is not much humidity in the atmosphere along with good rainfall & dry atmosphere. However, at the time of flowering, low temp. is essential like winter or monsoon.

However, hot & temperate climatic conditions are the habitat of the mango tree. However, temp. ranging between 24 to 30 °C is considered as the best suitable for mango cultivation for getting more fruit production.

Land Selection

As mango trees are large in size, so high-speed winds can ruin your mango cultivation. So, it is better to stay away from those areas where winds and cyclones keep coming. They can lead to loss of flower & mango fruits in your mango farm. Sometimes, heavy winds break the branches badly.

Take care that mango farming can be done over a wide range of soil, but it is beneficial if mangoes are cultivated in the more salty, alkaline, rocky, and water-filled land. So, mango cultivation in such type of land should be avoided for profitable tree farming business.

Planting Mangoes

Propagation Methods

For commercial mango farming, mangoes are mainly propagated by the grafting method like veneer grafting, epicotyl grafting, arching grafting etc. Good quality of mango trees can be prepared in less time by the Vinier and Softwood Grafting. It is suggested to learn the complete grafting techniques & then follow it if you're the new one in this. It will be more beneficial to you.

However, mangoes also can be propagated with the help of seed.

Mango Grafting Season

Planting mango should be done in the beginning of the rainy season (Monsoon season) because the risk of plant death is comparatively low in the rainy season. Rainwater helps the plant to fix faster. But, the best time for planting mango plant is from the month of July to August.

Planting of mango plant in the irrigation region should be done from the Feb. to March. Take while cultivating mangoes in the massive rainfall Zones. In such type of area, planting should be carried out at the end of rainy season.

Spacing

The spacing varies from 10 × 10 M to 13 × 13 M. In dry areas, the spacing should be 10 × 10 M because of less growth. While in heavy rainfall area and fertile soil, the spacing should be 13 × 13 M because of higher physical growth.

If your soil is infertile, then it should be supplemented at the time of planting onwards. For this, prepare a suitable pit. Then add about 25 kg of farmyard manure along with 3 kg of Super Phosphate & one kg of Potash.

However, the row spacing (about 10 m) along with tree spacing (around 5 m) is supposed to be the best mango tree plantation distance. With the help of this, you will be able to plant around 70 mango plants per acre, which is considered as avg mango plantation per acre.

Besides this, one can also go for ultra high-density mango plantation for more fruit production.

Irrigation in Mango Farming

Mango trees are vigorous in nature, so needs less water than other commercial fruit tree farming like BANANA FARMING. So, your mango farm should be irrigated at an interval of two to three days, in the first year of your mango farming business. But, when they start to bear fruit, at this stage, about two irrigation is essential.

Also, give water to those fruits bearing plant at an interval of 10 to 15 days. And stop irrigation after their full growth. Take care during the flowering stage. At this stage, stop giving them water. If irrigation does not stop, it may lead to bad quality mango fruit production.

How much water does a mango tree need? It depends on the type of soil, on which you are growing mango along with the growth of your mango tree. Give frequent water to your mango tree, after the plantation, at the time of flowering & fruit bearing stage.

Application of Manure and Fertilizer in Mango Farming

Your soil should be rich in some essential organic matter like Nitrogen, Potash, Phosphorous, Boron etc to prevent your fruit farm from some of the diseases. If your soil has any deficiency, then they should be supplemented at the time of soil preparation to earn higher production. So, for improvement of your soil; physically and chemically, application of cow dung manure, about 20 tonnes per hectare is beneficial.

For the first year of planting mango, supply about 50 gm of P_2O_5, 100 gm of Nitrogen, & 100 gm of K_2O. Add up this dose for the following years, one time for one year, up to ten years. After that, supply 500 gm of P_2O_5, 1 kg of Nitrogen, and 1 kg of K_2O to each mango plant. Apply these manure & fertilizers in the month of July to August.

Friends, application of Urea to the base of the mango tree at flowering stage will directly enhance the growth of the mango tree, yielding more amount of fruits. Apply manure & fertilizers for around ten years of planting the mango tree, each & every year after planting. And always try to supply them, particularly.Also, application of 40 kg cow dung manure along with 250 gm of Agozapirillum per each mango plant, is beneficial in enhancing the growth of plants.

Disease and Pest

The mango tree care is one of the key factors, which decide the production & profit of your mango tree farming. So, you have to care your mango farm for preventing the pest & diseases from your mango farm. Let us learn what are these mango diseases.

Mango Disease

- Powdery mildew

- Phytophthora fruit rot

- Anthracnose

- Bacterial black spot

- Mango malformation disease

- Apical bud necrosis

- Bacterial flower disease

- Stem-end rots are some common mango tree diseases that are found in mango farming business.

Presence of these diseases can be easily identified, but please consult your nearest horticulture for more details on the mango disease & their symptoms.

Mango Pest

- Mango shoot caterpillar,

- Fruit-piercing moths,

- Mango stem miner,

- Red-banded thrips,

- Fruit-spotting bug,

- Helopeltis,

- Red-banded mango caterpillar,

- Mango tip borer,

- Mango leafhopper,

- Mango seed weevil,

- Queensland fruit fly,

- Spiralling whitefly, etc are some commonly found pest in mango farming.

Make use of a particular pesticide for controlling certain mango pest.

Intercultural activities in Mango Farming

Intercropping

Mango tree starts to bear fruit at age of around five to six years. Up to that time, farmers have to wait for the first picking of mango. So, for farmers used to perform intercropping of any vegetable, guava, peach, papaya, legumes, plum, etc to earn extra income. By this, the extra space present in your mango farm will utilize perfectly. Depending on your climatic conditions, you can intercrop any crop in the unused space in your mango farm. But, take care that the Nutrition requirement & need of irrigation of the intercrop is different from the mango cultivation. Prepare your land on the requirement basis.

Other Intercultural activities

In mango farming, apart from intercropping, there are also some activities which are helpful in increasing the production amount. Here below are Intercultural Operation guide, perform them carefully.

1. Train your mango tree, so that your mango tree grows well.

2. Pruning mango trees should be done so that mango tree acquires desired shape.

3. Crossover branches should be discarded at pointer thickness.

4. Diseased & dried branches are to be discarded.

5. Mulching & controlling weeds with the help of weedicides is beneficial in enhancing fruit production.

6. Also, controlling mango pest & disease is enhance the mango fruit production effectively.

Harvesting Mangoes

Mango Harvesting is an interesting & important task. However, the mango production is mainly depended on the agroclimatic condition of your mango farming business. If climate supports a well, a good production of mango fruit can be obtained.

In mango cultivation, some types of mango start giving early production & some of the mangoes types used to late fruit bearing. Presently, with the help of various grafting method, it is available to get early production. Usually, grafted mango tree start to bear fruit from the fifth year onwards, while the mango tree grew with the help of seeds used to bear fruit at age of ten years.

Picking mangoes off a tree, it should be carried out on the basis of their maturity. If you want to sell them in the local market, then pick them at their maturity. And if you want to export your production, then pick them just before their maturity.

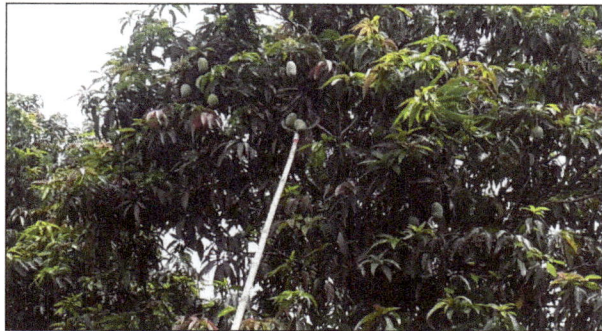

Yield

A good yield in mango farming business can be easily achieved with the help of good farm management skill, proper care & management of your mango farm. Controlling mango pest & disease is also a key factor in deciding the mango fruit production. However, with favorable climatic condition & proper grafting method, it is easy to increase mango production about 100 %. Timely application of suitable manure & fertilizers also enhances the mango fruit production as well.

Avg. around 10 tonnes of mango yield per acre can be easily obtained with the help of proper grafting method & suitable agroclimatic condition, increasing each & every year after the good establishment of your mango farm.

Banana

A banana is an edible fruit – botanically a berry – produced by several kinds of large herbaceous flowering plants in the genus Musa. In some countries, bananas used for cooking may be called "plantains", distinguishing them from dessert bananas. The fruit is variable in size, color, and firmness, but is usually elongated and curved, with soft flesh rich in starch covered with a rind, which may be green, yellow, red, purple, or brown when ripe. The fruits grow in clusters hanging from the top of the plant. Almost all modern edible seedless (parthenocarp) bananas come from two wild species – Musa acuminata and Musa balbisiana. The scientific names of most cultivated bananas are Musa acuminata, Musa balbisiana, and Musa × paradisiaca for the hybrid Musa acuminata × M. balbisiana, depending on their genomic constitution. The old scientific name for this hybrid, Musa sapientum, is no longer used.

Musa species are native to tropical Indomalaya and Australia, and are likely to have been first domesticated in Papua New Guinea. They are grown in 135 countries, primarily for their fruit, and to a lesser extent to make fiber, banana wine, and banana beer and as ornamental plants. The world's largest producers of bananas in 2017 were India and China, which together accounted for approximately 38% of total production.

Worldwide, there is no sharp distinction between "bananas" and "plantains". Especially in the Americas and Europe, "banana" usually refers to soft, sweet, dessert bananas, particularly those of the Cavendish group, which are the main exports from banana-growing countries. By contrast, Musa cultivars with firmer, starchier fruit are called "plantains". In other regions, such as Southeast Asia, many more kinds of banana are grown and eaten, so the binary distinction is not useful and is not made in local languages.

The term "banana" is also used as the common name for the plants that produce the fruit. This can extend to other members of the genus Musa, such as the scarlet banana (Musa coccinea), the pink banana (Musa velutina), and the Fe'i bananas. It can also refer to members of the genus Ensete, such as the snow banana (Ensete glaucum) and the economically important false banana (Ensete *ventricosum*). Both genera are in the banana family, Musaceae.

Female flowers have petals at the tip of the ovary

The banana plant is the largest herbaceous flowering plant. All the above-ground parts of a banana plant grow from a structure usually called a "corm". Plants are normally tall and fairly sturdy, and are often mistaken for trees, but what appears to be a trunk is actually a "false stem" or pseudostem. Bananas grow in a wide variety of soils, as long as the soil is at least 60 cm deep, has good drainage and is not compacted. The leaves of banana plants are composed of a "stalk" (petiole) and a blade (lamina). The base of the petiole widens to form a sheath; the tightly packed sheaths make up the pseudostem, which is all that supports the plant. The edges of the sheath meet when it is first produced, making it tubular. As new growth occurs in the centre of the pseudostem the edges are forced apart. Cultivated banana plants vary in height depending on the variety and growing conditions. Most are around 5 m (16 ft) tall, with a range from 'Dwarf Cavendish' plants at around 3 m (10 ft) to 'Gros Michel' at 7 m (23 ft) or more. Leaves are spirally arranged and may grow 2.7 metres (8.9 ft) long and 60 cm (2.0 ft) wide. They are easily torn by the wind, resulting in the familiar frond look.

Banana 'tree' showing fruit and inflorescence

When a banana plant is mature, the corm stops producing new leaves and begins to form a flower spike or inflorescence. A stem develops which grows up inside the pseudostem, carrying the immature inflorescence until eventually it emerges at the top. Each pseudostem normally produces a single inflorescence, also known as the "banana heart". (More are sometimes produced; an exceptional plant in the Philippines produced five). After fruiting, the pseudostem dies, but offshoots will normally have developed from the base, so that the plant as a whole is perennial. In the plantation system of cultivation, only one of the offshoots will be allowed to develop in order to maintain spacing. The inflorescence contains many bracts (sometimes incorrectly referred to as petals) between rows of flowers. The female flowers (which can develop into fruit) appear in rows further up the stem (closer to the leaves) from the rows of male flowers. The ovary is inferior, meaning that the tiny petals and other flower parts appear at the tip of the ovary.

Wild banana with flowers and stem growing in reverse direction

The banana fruits develop from the banana heart, in a large hanging cluster, made up of tiers (called "hands"), with up to 20 fruit to a tier. The hanging cluster is known as a bunch, comprising 3–20 tiers, or commercially as a "banana stem", and can weigh 30–50 kilograms (66–110 lb). Individual banana fruits (commonly known as a banana or "finger") average 125 grams (0.276 lb), of which approximately 75% is water and 25% dry matter (nutrient table, lower right).

Young banana plant

The fruit has been described as a "leathery berry". There is a protective outer layer (a peel or skin) with numerous long, thin strings (the phloem bundles), which run lengthwise between the skin and the edible inner portion. The inner part of the common yellow dessert variety can be split lengthwise into three sections that correspond to the inner portions of the three carpels by manually deforming the unopened fruit. In cultivated varieties, the seeds are diminished nearly to non-existence; their remnants are tiny black specks in the interior of the fruit.

Major Producers of Banana

When it comes to production, INDIA tops the list of banana producer all over the world. As banana is native of Northern Australia, Southeast Asia, & some tropical parts of India, produce widely by those countries because of the delicious taste of Banana fruit, is getting more popularity than other commercial fruit farming like Pomegranate Farming or Dragon Fruit Farming.

Here is the list of top 10 producers of banana all over the world:

1. India
2. China
3. Uganda
4. Philippines
5. Ecuador
6. Brazil
7. Indonesia
8. Colombia
9. Cameroon
10. Tanzania.

Factors affecting Production

Banana production; about more than 20 kg of banana fruit per each blossom of banana can be achieved if plant manages well & care properly. Because of this much banana yield per tree, it may result in the most profitable fruit trees to grow. So, it is a good thing to learn how can we can raise yield.

Here is a list of some factors that give their direct effect on the production of Banana Farming:

1. Varieties (cultivar) of Banana.
2. Climate Condition & Soil preparation.
3. Propagation & timely irrigation.
4. Timely application of suitable Manure and Fertilizers.
5. Care & management of your banana farm.
6. Post harvesting Care.

All these mentioned are key factors in deciding the production in Banana cultivation.

Varieties of Banana

Saba Banana, Grand Nain Banana & Cavendish Banana are the most popular cultivar of Banana, used mostly by the banana farmer to achieve higher yield per acre land. The main purpose of going for a hybrid is to produce optimum as possible. And select only high yielding & fast growing cultivar so that you became able to harvest them in short time.

Dwarf Cavendish, Karthali, Safed Velchi, Rasthali, Awak, Tanduk, Monthan, Red banana, grand nain, Nyali, Raja Abu, Robusta, Basara, Ardhapuri, Poovan, Karpurvalli, Nendran, Robusta, Hill Banana, Monthan, Karthali, Poovan, Emas, Nendran, Nangka and Rastali are some common varieties that are used by the Banana farmer for cultivating banana.

Select a cultivar, specific to your climate condition & is also high in yield.

Climate Condition for Banana Farming

If climate supports a well then you can easily obtain optimum production from it by selecting a high yielding cultivar. So, selection of banana cultivars is so based on the atmosphere. Since Banana plants are native to the tropical area, they grow well in a region with an atmospheric temp, ranging from 15 °C to 35 °C. Which state that, can grow well in warmer season with fair moisture & low wind.

As this crop a warm seasonal crop, excessive cold cause to crop injuries. So, always select a place, where the temp. does not fall under 15 °C. A normal temp about 18 °C to 30 °C is basically required for good development of the plants.

Soil Requirement for Banana Farming

Banana can be cultivated over a wide range of soil. However, a moistened soil with rich in organic matter & well drainage power along with 80 % loam, 70 % silt, & 45 % clay is considered as the best soil for the healthy vegetative growth of banana. It should slightly acidic in nature, having pH about 6.5 to 8. Do not farm it in lower pH soil because it causes to get more chance of disease in your crop like Panama disease. Also avoid cultivating banana in low laying areas, nutritionally deficient, ill-drained soil, sandy, and black cotton soils for higher production.

Organic matter like high nitrogen content, fair phosphorus with lots of potash ensure high production of Banana. If your soil is deficient of those organic matters, then it is a good idea to supplement them. Avoid planting bananas in water logging soil to prevent condition roots rot. However, this can be fixed by farming banana in raised gardens.

If your soil is deficient of any essential organic matters like Nitrogen, Potash, Phosphorus, etc then it is advised you to go for the soil test before starting farming Bananas.

Land Selection and its Preparation in Banana Farming

A proper site for your commercial Banana farming is also an important task as this will directly impact the overall production. Select a suitable place for your banana farm, which is rich in the Organic matter. Organic matter like high nitrogen content, fair phosphorus with lots of potash ensure high production.

Select an open place for your banana farming because they thrive their best in bright sunlight. Bring the selected soil into fine tilth form which can be easily achieved by 2 or 3 plowing with a tractor. But as they are a tree like herbs, a speedy wind may lead to breaking down of the plant. So, a monsoon with an avg. rainfall of 700 mm is good for the healthy vegetative growth of banana. Some varieties like 'Hill banana" are not able to give you more production at higher heights.

Growing Cowpea & Daincha, and burying them in the soil can help in increasing fertility of your soil as they are green manuring crops.

Here below is a guide about field preparation of different types of land:

- Garden lands: About two to four ploughings are enough for such type of soil.

- Wetlands: For such type of land, there is no need of ploughings.

- Padugai: About one to two ploughings are enough for such type of soil.

- Hill Banana: For farming bananas on the hills, clean the forest & construct a stone wall for your banana farm.

If your soil is deficient of organic matter, then it is a good thing to supplement those at the time of field preparation. However, application of farm yard manure is considered as the best one. So, for farming banana, an addition of farm yard manure about 45 tonnes per unit hectare land is enough. This will enhance soil fertility & also, the plant growth. Make sure that it mixed well with the uppermost soil.

Propagation in Banana Farming

Propagation in Banana farming mainly done with the help of Suckers & with the help of Tissue Culture. Mostly used to planting banana by suckers.

For planting bananas with the help of Suckers, suckers are sword well by using well-grown rhizome (a conical or spherical in shape) with active developing buds. They should be weighed not more than 700 gm for using them as propagating matter in the Banana cultivation.

Suckers or Tissue Culture

In banana farming, most of the people used to planting banana with the help of sucker because it is a low-cost method. While planting banana with the help of tissue culture seedling is a costlier method because, in this method, Banana seeds are grown by many operation activities after investing too much time.

As per my suggestion, for commercial banana farming, go for the tissue culture method of propagation. This is a costlier method but has lots of benefits over Suckers planting method. Due to variation in size & age of suckers, there is more chance of getting suckers infected of some common nematodes & pathogens. And also, a ununiform crop of banana which takes more time to become ready for harvesting. While in-vitro clonal propagation, banana plants are disease free, early yielding, healthy uniform in growth is observed.

Advantages of Tissue Culture

Here below is a list of some benefits of using Tissue culture plants in the Banana farming:

1. A disease & pest free seedling is observed.

2. Crop becomes ready for harvesting in very short time period.

3. Uniform growth of banana plant is observed with higher production.

4. Plant mature in short time than normal farming.

5. There is no staggered harvesting of bananas.

6. True to the type of mother plant under good management.

7. This method of farming make us able to farming bananas through out the complete year as easy availability of banana seedling.

8. With help of this, it is possible to obtain 2 consecutive raccoons in banana cultivation in short time, which directly decrease the cost of cultivation.

9. More plant productivity, about 95 to 99 % plants bears fruit bunches.

Planting in Banana Farming

Planting Season

With the help of tissue culture, it is possible to plant bananas during any time in a year. Try to avoid planting in excessive hot temp or coolant temp.

Banana Plantation Spacing

A good spacing in your commercial farming is also a key factor, which decides the production. So, for Banana plantation spacing, planting should be done with spacing 1.8 meters × 1.5 meter. This will make you able to plant more than 3600 banana plant per unit hectare land & is also assumed to be most economical and efficient spacing for commercial banana cultivation.

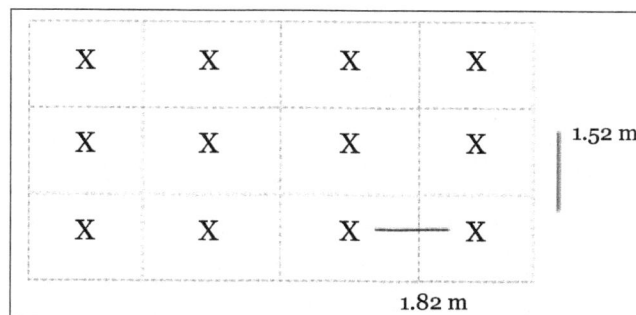

For high-density plantation or High-Density Planting, plant bananas at with spacing of 1.6 meter × 1.6 meters. But this may cause to lower production because of more struggle for sunlight between the plant. For higher production, it the recommended to plant them at the spacing of 2 M × 2.5 M. By this, you can easily raise, more than 2000 banana tree per unit hectare land & is also minimize the incidence chance of Sigatoka.

Planting Method

Planting method needs more care, from rest of the activities. So, plant them in a suitable way. For planting, removed polyethylene bags without disordering root ball of the plant. The plant it in the prepared hole keeping distance of 2 to 3 cm between the pseudo-stem. Take care while pressing the soil around the planted plant. Do not press it hard, as it causes to irregular growth of the plant. Also, avoid deep planting.

Irrigation in Banana Farming

A Banana crops require too much water for the healthy vegetative growth of banana fruit because they are a water loving plant. To obtain maximum productivity from it, give water too much to this commercial crop. And in monsoon, an avg. rainfall of 700 mm is good for the healthy vegetative growth of banana. In the critical condition like drought, deep irrigation should be provided to this crop.

Just after planting it on the field, a frequent irrigation is essential. Provide water sufficient enough for good development of plants. A banana tree did not bear any fruit bunch, if properly not irrigate. But, always try to avoid excessive watering, it will lead to roots rotten which finally results in low production in your crop.

Irrigation with the help of drip irrigation system more preferred over the traditional method. As drip irrigation is helpful in making effective use of water, manure as well as of fertilizer. It also lessens the chance of new emergence of weed. This also saves your precious & valuable time as it does not require any laborious work to give water with drip irrigation. It is recommended in those regions, where there is less availability of enough water.

Intercultural Activities and Weed Contol

Weeding is a common problem of all type of crops. Controlling it can lead to higher production in your crop. Here is some task to perform, to get rid of the weed.

Spraying, Glyphosate about 2 liters per unit hectare before plantation is helpful in controlling the weed. However, 2 manual weedlings are necessary to keep weed from the field.

Micronutrient Foliar Spray

To improve physiological, morphological, and the yield attributes, application of $FeSo_4$ (0.2%), H_3Bo_3 (0.1%), $CuSo_4$ (0.2%) along with $ZnSo_4$ (0.5%) as combined foiler spray has effective results. This micro nutrient spray by mixing up the following per 100 liters of water.

1. Zinc Sulphate (500 gm): Add 10 ml of sticker solution like Teepol in 500 gm $ZnSo_4$, before spraying it on the field.

2. Ferrom Sulphate (200 gm): Add 10 ml of sticker solution like Teepol in 200 gm $FeSo_4$, before spraying it on the field.

3. Copper Sulphate (200 gm): Add 10 ml of sticker solution like Teepol in 200 gm $CuSo_4$, before spraying it on the field.

Removal of the Male Buds

Remove male buds is beneficial for the healthy & vegetative growth of fruit. This also increases bunch weight via enhancing fruit growth.

Bunch Spray

Spraying of monocrotophos (about 0.2%), just after the emergence of fruits to avoid thrips. Thrips discolor the healthy fruit causes to became unattractive fruit. Take care that people are always looking for fresh & healthy fruit in the market. They always avoid buying such unattractive fruit, even at the too low rate.

Bunch Covering

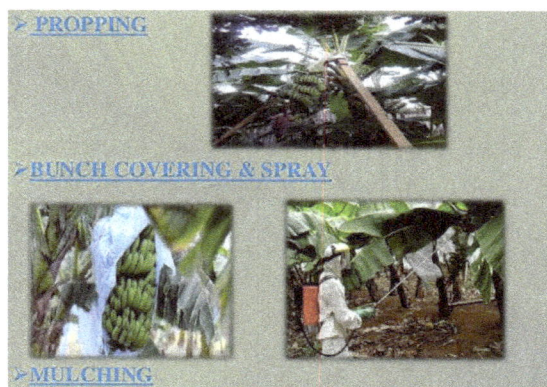

To obtain quality banana production, the fruit bunched should be covered by polyethylene or covering material to prevent it from direct sunlight. Covering these fruit bunches by using dried leaves is economically beneficially to the growers. Bunch covering also provides protection from dust, birds, insect and also spray residue. Covering with blue plastic is a good idea. This will help in raising the temp. for good development of fruit & finally the early maturity of fruits.

Dehandling of False Hands of the Bunch

Incomplete buds, which are incapable of producing quality fruit should be removed, just after the bloom. This will increase the weight of other fruit bunches. Mostly, they appear just above the false hand. So, removed them manually.

Propping

Fruit bunches are heavy in weight because of which plant lose their body balance & may fall on the field. To support them, propping of the tree should be done. Propped these overload tree with the help of 2 bamboo sticks, forming a triangle between them to the stem. This will also ensure the uniform growth of fruit bunches.

Application of Manure and Fertilizers in Banana Farming

Timely application of suitable manure & fertilizer is proper dose is essential in achieving a production. Organic matter like high nitrogen content, fair phosphorus with lots of potash ensure high production. So, your solid should be rich of these all organic matter. Any one can obtain a good yield, even there is nutrient deficient soil for this cultivation.

Make your soil rich in all essential nutrient, required for good production and application of farm yard manure the best one. So, give about 15 kg of farm yard manure, about 250 gm Nitrogen, about 75 gm Phosphorous, about 300 kg Potassium to each & every plant. If your soil has some nutrient deficiency, then they should be should be the supplement at the time of field preparation.

We can also provide these essential organic matters (Nitrogen, Phosphorous, & Potassium) in water with the help of drip irrigation system as it makes effective use of them as this gives direct water to the stem of the plant. However, the successful farmer used to give about 0.1 kg Phosphorous & 0.25 kg Potassium at the time of planting banana crops while the Nitrogen is given in 3 equal doses; at the time of vegetative growth, at the time of reproductive stage to continuously raise the productivity of the soil.

Pests and Diseases in Banana Farming

High-density planting in Bananas is more likely to be attacked by fungal diseases, Viral diseases also some of the insect (pests) which directly give their impact on quality & quantity production of fruit.

Pest

The common pests found in the commercial banana crop are listed below:

1. Nematodes

2. Rhizome weevil

3. Fruit scarring battle

4. Pseudostem weevil

5. Thrips and Aphids.

Fungal Diseases

The common fungal diseases that attacks on the commercial banana crop are listed below:

1. Panama wilt

2. Head rot and Sigatoka leaf spot.

Viral Diseases

The common viral diseases that attacks on the commercial banana crop are listed below:

1. Banana Bract Mosaic Virus

2. Banana Streak Virus

3. Banana Mosaic Virus and Banana Bunchy Top Virus.

Controlling Measure

For symptoms & controlling measures, please consult nearest a horticulture department as they are good source of reliable information for your local region.

Banana Harvesting

Timely banana harvesting leads to higher production of fruit. The duration for banana growth totally depends on the cultivars & climate condition of your surroundings. After about 250 to 360 days (normally). But harvest this delicious fruit, when they reach their maturity. At this stage, quality & quantity fruit can be obtained which is high in market demands. Cultivate only those verities which are high yielding & fast growing. It will make you able to earn a big profit in the very short time period with high production.

For large scale commercial banana farm, tissue culture planting helps in collecting two raccoons of banana in very short period. Harvest them & rip them at home. If you, not a good market for selling these fruit, then export them to other markets. Usually, fruits are harvested after about three to four month of opening the 1st hand.

Removed carefully & collect these fruit bunches in a basket at a good collecting place. Keep them in sunlight since it accelerates the ripening and qualifying of fruits. For local consumption, hands are usually left on fruit stalks & sold to retailers.

Yield in the Banana Farming

By this crop, in a time period of two & half-years (26 to 32 months), you became able to produce as high as 100 tonnes of fruits per unit hectare land with the help of tissue culture technique. With help of tissue culture, it is also possible to obtain more raccoons with the help of Fertigation & drip irrigation system.

Under the use of combined drip irrigation with, a yield of Banana as high as 100 Tones per hector can be easily obtained with the help of tissue culture technique, even similar yield in the ratoon crops can be achieved if the crop is managed well.

Post Harvest Management

Here below are some suggestion to perform after harvesting of fruit:

1. Discard over mature and injured fruits.

2. Clean fruit below running water by using dil sodium hypochlorite sol.

3. Remove latex, if present by treating fruit with thiabendazole.

4. Short out on the basis of size as Grade A, B, C.

5. For local market, use lorries for transportation to minimize your cost Or pack.

6. Pack it in suitable box & store fruit in cool place to maintain fruit quality.

Apple

Apple (Malus pumila) is an important temperate fruit. Apples are mostly consumed fresh but a small part of the production is processed in to juices, jellies, canned slices and other items.

Climate

The apple is a temperate fruit crop. The average summer temperature should be around 21-24 °C during active growth period. Apple succeeds best in regions where the trees experience uninterrupted rest in winter and abundant sunshine for good colour development. It can be grown at an altitude of 1500- 2700 m above the sea level. Well-distributed rainfall of 1000-1250 mm throughout the growing season is most favourable for optimum growth and fruitfulness of apple trees.

Soil

Apples grow best on a well-drained, loam soils having a depth of 45 cm and a pH range of pH 5.5-6.5. The soil should be free from hard substrata and water-logged conditions. Soils with heavy clay or compact subsoil are to be avoided.

Propagation

- Grafting: Apples are propagated by several methods viz.; whip, tongue, cleft and roots grafting. Tongue and cleft grafting at 10-15 cm above the collar during February-March gives the best results. Usually grafting is done at the end of winter.

- Budding: Apples are mostly propagated by shield budding, which gives a high percentage of success. In shield budding a single bud along with a shield piece of stem is cut along with the scion and inserted beneath the rind of the rootstock through a 'T' shaped incision during active growth period. Budding is done when the buds are fully formed during summer.

- Rootstocks: Most of the apple plants are grafted or budded on seedling of wild crab apple. The seedling rootstocks obtained from the seeds of diploid cultivars like Golden Delicious, Yellow Newton, Wealthy, Macintosh and Granny Smith also can be used. High density planting is done using dwarfing rootstocks (M9, M4, M7 and M106).

Planting

The planting distance varies according to variety and the fertility level of the soil. The main consideration in planting trees is planting of sufficient pollinators to ensure effective pollination. Usually one pollinator tree is needed for two to three large trees planted at 10 m distance or one row pollinator for two rows of main cultivar. For high density planting the pollinator tree is planted after every sixth tree in a row.

The most widely used planting system is the square system. In this system, the pollinators are planted after every sixth or ninth tree. The other popular system of planting is the rectangular system. In hilly areas the apple orchards are established by planting the trees on the contours so as to prevent soil erosion and reduce run off.

The average number of plants in an area of one ha. can range between 200 to 1250. Four different categories of planting density are followed viz. low (less than 250 plants/ha.), moderate (250-500 plants/ha.), high (500-1250 plants/ha.) and ultra high density (more than 1250 plants /ha.). The combination of rootstock and scion variety determines the plant spacing and planting density/unit area.

Time and method of Planting

Planting is usually done in the month of January and February.

Pits measuring 60 cm are dug two weeks before planting. The pits are filled with good loamy soil and organic matter. Planting is done in the centre of the pit by scooping the soil and placing the soil ball keeping the roots intact. Loose soil is filled up in the remaining area and lightly pressed to remove air gaps. The seedlings are staked and watered immediately.

Irrigation

Apple trees are particularly sensitive to low soil moisture. Water stress during the growing season reduces number and size of fruits, and increases June drop. Success of apple largely depends on uniform distribution of rain during the year in case of dry spells during the critical periods supplementary irrigation should be provided. Water stress conditions results in poor fruit set, heavy fruit drop, low production and poor quality. The most critical periods of water requirement are April- August and peak water requirement is after fruit set. Normally the orchards are irrigated immediately after manuring in the month of December-January. During the summer periods, the crop is irrigated at an interval of 7-10 days. After the fruit setting stage the crop is irrigated at weekly intervals. Application of water during the fortnight preceding harvest markedly improves the fruit colour. Thereafter till the onset of dormancy, irrigation is given at an interval of 3-4 weeks.

Pruning

Pruning is one of the most important practice which promotes plant vigour and productivity. Pruning is done with a view to divert the sap flow towards the fruiting branches and to force the plants to bear more fruits or to induce vigorous vegetative growth. During pruning, weak-growing and diseased branches are removed from the tree. Usually the trees are pruned every year in the month of December-January. The systems of pruning adopted in apple cultivation are as follows:

- Established Spur System: Objective of this pruning is to develop permanent fruit spurs for production of fruits. To ensure formation of spurs on the laterals the central leader is cut back every year along with the strong erect laterals near the central leader. This leads to wide angled vigorous laterals for formation of spurs.

- Regulated System: Regulated pruning is practiced generally on apple cultivars growing on semi-dwarfing and vigorous rootstocks. Before planting, the central leader of the tree is cut back at 75 cm on which three well- placed primary branches are allowed to grow. In bearing trees, the growth of leader and strong laterals are encouraged by pruning weak and crowded branches.

- Renewal System: In vigorous cultivars instead of developing permanent spurs, the objective is to encourage continuous growth of new healthy shoots, spurs and branches every year. A part of the tree is pruned every year to produce fruits in the following year on the new shoot growth, while the unpruned parts produces fruit buds.

Thinning of Fruit

Thinning is one of the major techniques employed to regulate fruit quality. In apples, heavy bearing not only results in small-sized poor quality fruits but also sets in alternate bearing cycle. Judicious thinning done at the proper stage of fruit development can regulate cropping and improve fruit size and quality. Since manual thinning is cumbersome and expensive, chemical thinning is employed.

Chemical thinners should not be applied in very hot and dry conditions as it adversely affects the absorption. Spraying should be done thoroughly to cover the entire canopy. Sometimes chemical thinning follow calcium deficiency therefore adequate calcium nutrition should be supplemented after thinning.

It is desirable to retain one fruit for every 40 leaves. This spaces the fruit at about 15-20 cm apart and there will be only one fruit per spur.

Manuring and Fertilization

Farmyard manure @ 10 kg./ year age of tree is applied along with other fertilizers. The fertilizer dose depends upon the fertility of soil and amount of organic manure applied to the crop. Generally, application of 350 g N, 175 g P_2O_5 and 350 g K_2O per plant per year in split doses is recommended for fully-grown bearing trees. On some trees deficiency of zinc, boron, manganese and calcium may be observed which is corrected with the application of appropriate chemicals through foliage spray.

Plant Protection

Insect Pests : The insect pests mostly observed are San Jose Scale (*Quadraspidiotus perniciosus*), white scale (*Pseudoulacaspis* sp.), wooly apple aphid (*Eriosoma lanigerum*), blossom thrips (*Thrips rhopalantennalis*) etc. Planting of resistant rootstocks, suitable intercultural operations and spraying with chloropyriphos, fenitrothion, carbaryl etc. have been found to be effective in controlling the pests.

Diseases : The main diseases reported are collar rot (*Phytophthora cactorum*), apple scab (*Venturia inaequalis*), sclerotius blight (*Sclerotium rolfsii*), crown gall (*Agrobacterium tumefaciens*), cankers, die-back diseases etc. Plants resistant to the diseases should be used for cultivation. The infected plant parts need to be destroyed. Application of copper oxychloride, carbendazim, mancozeb and other fungicides have been found to be effective in controlling the diseases.

Disorders : In apple, there are three distinct fruit drops.

- Early drop resulting from unpollinated or unfertilized blossoms.
- June drop (due to moisture stress and fruit competition) and
- Pre-harvest drop. Pre-harvest drop can be controlled by spraying NAA @ 10 ppm. (1 ml. of Planofix dissolved in 4.5 l. of water) about a week before the expected drop.

Harvesting

Normally the apples are ready for harvest from September-October except in the Nilgiris where the season is from April to July. The fruits mature within 130-150 days after the full bloom stage depending upon the variety grown. The ripening of fruits is associated with the change in colour, texture, quality and the development of the characteristic flavour. The fruits at the time of harvest should be uniform, firm and crisp. The colour of the skin at maturity ranges from yellow-red depending on the variety. However, the optimum time of harvest depends on fruit quality and intended period of storage. Due to the introduction of dwarf rootstock hand picking is recommended as it reduces bruising due to fruit fall during mechanical harvesting. Yield - The apple tree starts bearing from 4 year onwards. Depending on variety and season, a well managed apple orchard yields on an average 10-20 kg/tree/year.

Orange

Like Banana, Apple & Mango, Orange is also one of the topmost fruit, that is cultivated widely in every corner of the world. Basically, oranges are citrus fruits, frequently eaten by the people because of orange nutritional value. Orange fruits are the excellent source of vitamins, particularly the vitamin C. And, Orange juice acquire an important place in the regular diet of many people. So, friends, growing orange tree is a great idea to begin a new tree farming business plan.

The scientific name of Orange is Citrus reticulata & the botanical name of orange is Citrus aurantium.

Oranges products have great demand in Internation Market because of its extensive scope of uses & Benefits. This delicious fruit can be eaten as a fresh or in form of syrup, juice, jam, squash, or any other orange products. Note that, orange fruits are the principal source of citric acid, peel oil, and cosmetics.

Top Orange Producing Countries

Here below is a list of top 10 countries, which top the orange production list:

1. Brazil

2. USA

3. China

4. India

5. Mexico

6. Spain

7. Egypt

8. Turkey

9. Italy

10. South Africa.

Climate Condition for Orange Farming

Likewise Mango Farming and Banana Farming, Oranges are also cultivated widely as they can grow well in the tropical region as well as in sub-tropical regions up to a height of 1500 m. However, a dry climatic condition is favorable for profitable orange cultivation for obtaining higher oranges production.

A soil temp. (about room temp: 26 °C) is beneficial in enhancing the growth of plant root as well plant growth. However, farming oranges in an arid and drier area with good hot summer and moderate rainfall, about 100 to 200 cm is considered as the best and most favorable climatic conditions for the excellent growth of orange plants.

Take care that orange trees are very sensitive to high humid and frost condition and such climatic condition in orange farming invites numbers of many diseases and harmful pest, insect, etc. Also, note that hot winds during summer lead to dropping down of orange flower and budding fruits. So, for obtaining high production of oranges, it is a good idea to keep maintaining the favorable temp in your farm during the growth period.

Soil Requirement for Orange Farming

Oranges can be cultivated on a wide variety of soil like sandy loam to loamy soil, alluvial, red to black soils with too much clay. However, a soil with high fertility, good drainage power, lime-free soil with good salt absorption power and light loam is considered as the best & most favorable for the orange cultivation.

As soil ph for citrus trees is about 6.0 to 7.0 and oranges are also a citrus tree, it is smartness to cultivate them on soil, which has pH ranging between 5. 8 to 7.3 for best production. However, it is a good idea to go for the soil test, at least once before starting any commercial farming, to know that is your soil is suitable or not.

Land Selection and its Preparation for Orange Farming

Select a site for your orange farm, from where there is a full availability of transport for transporting your oranges in the market. Also, it should not be too cold during the winter and monsoon season because this invites numbers of pest and disease to your farm. It is also suggested to give windbreaks on sides of your farm from heavy winds since hot and heavy winds during summer lead to dropping down of orange flower and budding fruits. Growing Jamun (Plum), Mulberry trees, Eucalyptus, etc on sides is a good idea for providing windbreaks.

Prepare your soil, favorable for orange farming like managing ph about 5.8 to 7.3, make rich your soil in all essential organic matters for producing more fruits. If your soil has any nutrient deficiency, then it should be supplemented at the time of soil preparation and remover the previous crop's weeds, if present.

Propagation in Orange Farming

Propagation Method

An Orange tree can be reproduced by seed or with the help of vegetative propagation, by T-budding.

Learn how to grow an orange tree from a seed in few steps:

1. Prepare a pot, having sterilized pot soil.

2. Plant your orange seed into this pot at about 1/2 inch depth in pot soil.

3. Cover this pot by using any wrapping material.

4. Keep this pot in a warm place and give some time to grow.

5. When seedling appears, move your pot to the sunny and open area so that it can grow well

6. Water it regularly.

However, planting oranges with the help of budding are fast growing & high yielding in nature.

Growing Oranges - Planting Season

Planting oranges should be done in the beginning of the rainy season (Monsoon season) because the risk of plant death is comparatively low in the rainy season. Rainwater helps the plant to fix faster. But, the best time for planting orange plant is from July to August.

Spacing

Prepare a pit, having size 80 cm × 80 cm × 80 cm. Add about 10 kg of rotten cow dung manure to the topsoil of the pit. Plant the Oranges, keeping a distance of 7 meters between the rows and

7 meters apart. However, the spacing in orange cultivation depends on the type of soil, method and the cultivars, selected for the cultivation.

Friends, for high-density orange planting, this spacing may be less. With less spacing, it is able to plant about 120 plants per one acre. Watering should be done, just after the planting.

Irrigation in Orange Farming

Friends, Orange plant require more water as like the Banana Plant. for the repetitive growth of fruit plant. However, irrigation in your orange farm depends on many factors like agro-climatic condition, plant age, and soil type. Give water your orange trees at a regular interval like at a week in the summer season and at 10 days from monsoon to winter. However, first irrigation should be done, just after the orange plantation.

Avoid waterlogging the tree, because orange trees are very sensitive to the water stagnation. A fully grown Orange plant needs minimum 20 watering in a particular year i.e 1,400 mm water. However, farming with drip irrigation system is the best and beneficial in any commercial crops.

Manure and Fertilizers in Orange Farming

Basically, an orange plant needs Nitrogen, Potassium, Phosphate along with some Micro-nutrient Magnesium, Zinc, Iron, etc for good development of vegetative plant growth. Keeping orange tree full of organic manure is helpful to the orange plant in bearing healthy and delicious oranges. So, here below is a guide for application of manure and fertilizer in orange farming for good fruit production.

Orange Plant Age	Year-wise Fertilizer dose (gm per plant)		
	N	P	K
1st Year	150	50	25
2nd Year	300	100	50
3rd Year	450	150	75
4th Year & above	600	200	100

Note that timely application of suitable manure and fertilizer play an important role in enhancing Orange fruit production. You can make use of any best fertilizer for orange trees in your farm

Orange Tree Pests and Diseases

The Orange tree care is one of the key factors, which decide the production & profit of your orange farming business. Care of orange tree is mainly done for preventing the disease and pest of the orange tree.

Orange Disease

Gummosis, twig blight, damping off, root and collar rot are some main disease, that are commonly observed in the orange farming business. For controlling the Orange tree disease in your farm, depending on your orange plant infection, spray Bavistin, Ridomil MZ 72, Benomyl etc on the affected orange plants.

Orange Pests

Snails, Citrus Red Mites, Citrus Bud Mite, Brown Soft Scale, Citrus leafminer, Citrus Thrips, Orangedog Caterpillars, Mealybugs, Citrus psylla, Citrus Whitefly, Fruit fly, Blackfly, Bark eating caterpillar, Aphids etc.. are some common insect and pest, that are observed on an orange tree.

Why Disease and Pest Control in Orange Farming?

Presence of all these pests and disease in your farm are the main cause beside the poor quality of oranges and also the low fruit production. So, they should be controlled as soon as possible. Liquid Pesticide have effective results for controlling pests and diseases in an orange cultivation business.

For controlling the Orange tree disease in your farm, spray Bavistin, Ridomil MZ 72, Benomyl etc on the affected orange plants, depending on your orange plant infection. While for controlling Pest, spraying quinalphos, dimethoate, monocrotophos, phosphamidon, phosalone, etc insecticides, depending on your orange plant infection. Timely application of suitable fertilizer in your orange farm is beneficial in controlling these pests and diseases and also helps in enhancing the fruit production in your fruit farming business.

Orange Harvesting

Generally, an orange plant becomes mature at the age of one year. After which they start flowering and budding, but not in proper amount or less no. of fruit are obtained in the first harvesting.

Usually, Depending on the types of orange cultivated, from the 4th year onward, you will be able to harvesting oranges. And an orange tree used to give fruit for about 25 yrs in an economic life. Collect those tasty and delicious oranges at three different intervals, during a year. Harvest them in autumn, summer, and in the rainy season.

For meeting a higher amount of profit, take care in picking oranges. They should be picked, when are in their best size and also ripen completely. Attractive color, bigger size, a good sugar content and well acid ratio are main factors, which yields a higher value of fruit in the market.

You can make use of orange harvesting machine for picking oranges, fresh and healthy as orange harvesting is a time-consuming process and require more laborious work.

Yield

After about four to five, year, you will be able to harvest these tasty oranges. However, the no's of fruit per tree in your fruit farming business is less at this stage, hence less is the yield also. However, 20 to 35 tonnes of oranges can be easily obtained, each and every year. This may high or low, depending on the types of oranges cultivated, soil fertility, agroclimatic condition, orange tree care and management.

In commercial orange cultivation business, about 50 no's of fruit can be obtained per each tree while after the good establishment of your orange farming business, about 500 no's of fruit per tree can be easily obtained.

Fruit Farming

Fruit farming is the growing of fruit crops, including nuts, primarily for use as human food.

The subject of fruit and nut production deals with intensive culture of perennial plants, the fruits of which have economic significance (a nut is a fruit, botanically). It is one part of the broad subject of horticulture, which also encompasses vegetable growing and production of ornamentals and flowers.

Botanists define a fruit in broad terms as the fleshy or dry ripened ovary surrounding the seed of a plant. A pomologist, or specialist in the science and practice of fruit growing, defines it somewhat more narrowly as the fleshy edible part of a perennial plant associated with development of the flower. A nut is any seed or fruit consisting of a kernel, usually oily, surrounded by a hard or brittle shell. Most edible nuts—e.g., almond, walnut, cashew, pecan, pistachio, etc.—are well known as dessert nuts. Not all nuts are edible. Some, used as sources of oil or fat, may be regarded as oil seeds; others are used for ornament. The botanical definition of a nut, based on features of form and structure (morphology), is more restrictive: a hard, dry, one-celled, one-seeded fruit that does not split open at maturity. Among the nuts that fit both the botanical and popular conception are the acorn, chestnut, and filbert; other so-called nuts may be botanically a seed (Brazil nut), a legume (peanut [groundnut]), or a drupe (almond and coconut).

Improvements in technology and consolidation of the fruit and nut industries in the most favoured climates of the world have been responsible for a steady increase in yield. Thus, the total acreage or number of plants devoted to various fruit and nut crops has dropped, remained about the same, or not risen in proportion to the increase in the respective crop production.

Although fruit- and nut-growing enterprises cover great ranges of climates and plant materials, their technologies have many common problems and practices.

The most significant of these are discussed below:

The Variety: its Propagation and Improvement

The first step in establishing a fruit- or nut-growing industry is the selection of individual plants with high productivity and a superior product. Such an individual is a horticultural variety. If it is multiplied vegetatively from rooted cuttings, from root pieces that throw shoots, or by graftage, each plant in the group (called a clone) that results is identical with the others. Nearly all commercially important perennial fruit and nut crops are clonally propagated; i.e., their varieties are multiplied vegetatively by one means or another. Some nut crops, such as the wild pecan, cashew, black walnut, hickory, and chestnut still come from trees that grow at random from seed; hence, character and quality tend to vary.

Many important varieties of fruit plants were selected generations ago. The Sultanina (Thompson Seedless) grape, the Lob Injir (Calimyrna) fig, and the Gros Michel banana have obscure origins; planted by the millions since selection, each specimen is actually a vegetative continuation of the selected individual growing on an independent root system. But regardless of the age of a fruit-growing industry, or the perfection of some of the selected varieties, a continuing search for new varieties is essential. There is always room for improvement in climatic adaptability, in insect and disease resistance, and in the solution of special horticultural or marketing problems. In fact, government experiment stations over the world now stress scientific breeding for improvement of market quality and yield of key fruit and nut crops.

Gros Michel banana Gros Michel banana trees.

Not only are varietal selection and improvement a continuing need but so also is the maintenance of existing varieties. Although an improved vegetative mutation of a variety is exceptional, the opportunities for accidental multiplication of degenerate (low-quality) mutants increase in proportion to the number of specimens of the variety. As a result, care is taken to propagate a clone only from superior individuals, and in the case of citrus, where mutation is especially common, further precautions are necessary. There are, of course, occasional mutations that may greatly improve a variety and these are sought, selected, and propagated.

Vegetative propagation technique varies with the individual fruit plant. Date, banana, and pineapple are multiplied by use of offshoots or suckers. Grape, fig, olive, currant, and blueberry are usually propagated from cuttings. Strawberry and black raspberry reproduce vegetatively by special organs—the former by stolons or runners, the latter by cane tip rooting or layering. Many kinds of fruit trees must be grafted or budded on especially grown rootstocks because the species to be multiplied does not root itself easily; apple, pear, peach, mango, and citrus are examples of this group. Many nut trees have a single taproot with but few branching roots, necessitating a deep hole and special care in transplanting.

Today's trend is toward a smaller tree in most fruit crops, particularly the apple and pear, and toward closer planting in hedgerow style, with carefully regulated fertilization and irrigation. This increases production per acre, lowers labour cost, increases early yields, and facilitates access in maintenance and harvesting. This approach, in fact, has been used for decades in Europe. Labour is the largest element of cost in fruit and nut production. Every means is exploited to reduce, facilitate, or eliminate hand labour.

With most fruit species a period of one to two years intervenes between the time a cutting is rooted and the time the plant is ready for setting in the field, or between graftage or budding and field planting. During this interval the plants remain in a nursery where they can be given intensive culture in rows. Pineapple and banana planting materials, however, do not require nursery care before field planting.

In choosing fruit varieties, the grower must (1) recognize the relative adaptabilities of available varieties to the climatic and soil conditions of his farm and (2) select a group that satisfies both his management needs and the market demands from those best adapted to his conditions. For instance, an apple producer in the northeastern U.S. may raise four varieties: Milton, McIntosh Red, Red Delicious, and Rome Beauty. The main harvest seasons for these succeed each other at two-week intervals; this helps him extend the harvest period and make efficient use of his labour. The first two varieties cross-fertilize satisfactorily, as do the last two. The first of these varieties is usually marketed without storage, while the storage seasons of the others are of increasing length. This helps the grower to extend his marketing period.

Cultivation

Site Selection

The site of a fruit-growing enterprise is as significant in determining its success as the varieties grown. In fact, variety and site together set a ceiling on the productivity and profit that can be realized under the best management. In most developed fruit regions microclimatic conditions (climate at plant height, as influenced by slight differences in soil, soil covering, and elevation) and soil conditions are the two components of a site that determine its desirability for a fruit-growing enterprise. Sometimes (particularly with highly perishable fruits) transportation to market must also be considered.

Local conditions at a site that expose it to unusual frost hazard are as detrimental to citrus in Florida as they are to peach trees in New Zealand and apple trees in the south of England. In regions and sites where temperatures during the season may drop no more than a few degrees below freezing, artificial frost protection is sometimes used. This is accomplished by open-flame burning (petroleum bricks, logs, etc.) or heating of metal objects with oil, gas, propane, electricity, etc. (stones or stacks that radiate heat). Another technique is the spraying of water on plants (e.g., strawberries) as long as the temperature is below freezing.

For highest productivity, most fruit trees must root extensively to a depth of three feet (one metre) or more. Heavy subsoil or other conditions causing imperfect internal drainage may result in shallow, weak root systems that do not take water and nutrients efficiently from the soil. In semi-arid and arid regions, accumulation of saline soils in a subsurface layer sometimes limits rooting of fruit trees, causes abnormal foliar symptoms, and reduces yields. Tiling and surface ditching help decrease water accumulation in poorly drained subsoils and reduce wet spots in otherwise satisfactory sites. Special control of irrigation procedures and periodic leaching may alleviate the worst salt effects in saline soils. Choice of tolerant species, varieties, and rootstocks may make fruit growing economical on imperfectly drained or mildly saline sites, though plants rarely perform as well as they do on sites free from these difficulties. Coconuts, however, tolerate saline soil conditions near tropical saltwater coasts.

Once selected, a site is cleared, levelled and cultivated. Then drainage, irrigation, and road systems are installed as required. In rolling or sloping terrain, where contour planting is needed to control erosion and conserve moisture, the locations of the plant or row positions are determined by the contour terraces and waterways established. In old lands, nematode or other pest populations make fumigation necessary before planting. In some problem California soils, giant plows and treaded tractors turn the soil to depths of three to six feet (one to two metres). In very infertile sites, or sites where the physical condition of the surface soil is poor, it may be helpful to grow a succession of leguminous cover crops for a year or more before planting and/or apply a fertilizer containing major fertilizer elements (nitrogen, potassium, phosphorus, calcium, sulfur) and all or certain trace elements (iron, manganese, boron, zinc, copper, molybdenum) and lime, based on a soil test.

Planting and Spacing Systems

Growth, flowering habits, and light requirements on the one hand, and management problems on the other, determine the most satisfactory planting plan for a fruit- and nut-growing enterprise. There is a trend toward use of dwarfing stocks, growth control chemicals, or closer planting and training, or all of them to get the highest yields and best operation efficiency possible on a unit of ground.

Low-growing crops such as strawberry and pineapple are usually managed in beds containing several rows, or in less formal matted rows. In an acre of strawberries, 200,000 or more plants may occupy the matted rows. A pineapple plantation with two-row beds, having plants one foot (0.3 metre) apart in rows two feet (0.6 metre) apart totals 15,000 to 18,000 plants per acre (37,000 to 44,000 per hectare). With such dense populations, intense competition for light, water, and nutrients causes smaller average fruit size. Nevertheless, the total yield per unit of land is usually greater than it would be with lower plant numbers.

The spacing of grapevines along a trellis row and of trees planted in hedgerows involves the same group of problems. Maximum vineyard production frequently results with vine distances of eight to nine feet (2.4 to 2.7 metres; 600 ± per acre [1,500 per hectare]). The trend for peach trees and spur-type apple strains is hedgerows 14 feet (4.2 metres) apart or closer, in rows 18 to 20 feet (5.4 to 6 metres) apart.

With those species and varieties that require cross-pollination by insects, the planting plan must take those special needs into account. This is a problem with apple, pear, plum, and sweet cherry orchards. At least two varieties that cross-fertilize successfully must be planted in association with each other.

Training and Pruning

Pruning is the removal of parts of a plant to influence growth and fruitfulness. It is an important fruit-growing practice. Primary attention is given to form in the first few years after fruit trees or vines are planted. Form influences strength and longevity of the mature plant as well as efficiency of other fruit-growing practices; pruning for form is called training. As the plant approaches maximum fruitfulness and fills its allotted space, maintenance pruning for various purposes becomes increasingly important.

The grape may be trained following one of two systems: (1) spur system, cutting growth of the previous season (canes) to short spurs, (2) long-cane system, permitting canes to remain relatively long. Whether a spur or long-cane system is followed depends on the flowering habit of the variety. Relatively small trees that respond favourably to severe annual pruning (e.g., the peach and Kadota fig) are usually trained to create an open-centred tree with a scaffold of four or five main branches that originate on a short trunk and branch a number of times to provide fruiting wood.

Annual renewal pruning can be reasonably efficient under these circumstances. Larger trees that do not respond favourably to heavy annual pruning are trained best to a system that encourages the main leader branch to grow erect to a height of eight to 10 feet (2.4–3.0 metres), with four or five main lateral branches at intervals on its sides forming the scaffold that carries fruiting wood up and out; this is called a modified leader system. The central leader type of tree, with one main leader up through the centre and many side branches, is common for pear and apple planted in hedgerows, and possibly for other fruits and nuts as the close-planted hedgerow system is more widely adopted.

The principal reasons for maintenance pruning are: (1) to permit efficient spraying and harvesting operations, (2) to maintain satisfactory light exposure for most of the leaves, and (3) to create a satisfactory balance between flowering and leaf surface.

To reduce hand labour costs, larger commercial fruit growers use machine pruning on many types of fruits. Peach, apple, pear, and other fruits usually planted in hedgerows are mowed across the top and sides by machine, then thinned out as needed by a follow-up crew using pneumatic clippers and hand-powered saws, operating from hydraulically manipulated scaffolds or lifts of various types.

Soil Management, Irrigation and Fertilization

Soil Management

Two soil management practices (1) clean cultivation and chemical weed control or both and (2) permanent sod culture, illustrate contrasting purposes and effects. In clean cultivation or chemical weed control, the surface soil is stirred periodically throughout the year or a herbicide is used to kill vegetation that competes for nutrients, water, and light. Stirring increases the decomposition rate of soil organic matter and thereby releases nitrogen and other nutrients for use by the fruit crop. It may also provide some improvement in water penetration. On the other hand, laying bare the soil surface exposes it to erosion; destruction of organic matter eventually lowers fertility and causes soil structure to change from loose and friable to tight and compacted. Though sod culture minimizes the destructive processes and may permit a modest increase in fertility, the sod itself competes with fruit plants for water and nutrients and may even compete for light. As a result, permanent sod culture is practical only with tree crops that are normally rather low in vegetation, such as apple, pear, sweet cherry, nuts, and mango. Competition from established sod may be detrimental to vigorously growing fruit plants like grape, peach, and raspberry unless adequate fertilizer and water are supplied.

Because each of these soil management systems has advantages and disadvantages, modifying or complementary practices are often used; for example, cover cropping, mulching, and chemical control of vegetation with or without strip sod in the row middles. In fact, the trend is toward mowed sod middles with strip chemical control under the trees and with overhead sprinklers during hot dry weather. Sprinklers not only provide water but tend to cool the plants and give fruit of better market quality without aggravating diseases. Cultivation combined with winter cover cropping has been used widely in grape, peach, cherry, bush fruit, and citrus plantings, as well as with other species. Mulching is the addition of undecomposed plant materials such as straw, hay, or processors' refuse to the soil under the plants. In orchards, mulching materials are most often applied under trees maintained in permanent sod. Strip in-row chemical control of vegetation in commercial fruit plantings has almost taken over as an economical and sound practice.

Irrigation

In semi-arid and arid regions, irrigation is necessary. Probably the maximum demand occurs in date gardens, because they expose a large leaf surface the year around under conditions of high evaporation and practically no rainfall. Irrigation in humid climates is generally being provided increasingly during extended dry periods that occur at one time or another during most growing seasons. For example, large acreages of banana are irrigated on coastal lowlands of the torrid tropics where annual rainfall exceeds 60 inches (1,500 millimetres).

Fertilization

Needs of perennial fruit plants for fertilizers depend on the natural fertility of the soil supporting them and on their individual requirements. Of the essential elements, supplemental nitrogen is almost always needed; potassium supplements may be needed, even in some desert areas. Although strawberry, grape, peach, and a few other fruits have responded favourably to phosphorus, and although its application has been recommended, the phosphorus requirement of woody plants is low and deficiency is rather rare. Calcium deficiency may be more common than realized; lime is often desirable to reduce soil acidity and because of other indirect benefits. Inadequate magnesium in the soil has been noted by workers studying a wide range of fruit species. Of the trace elements, zinc, iron, and boron are most likely to be deficient, but copper, manganese, and molybdenum deficiencies also are being reported for some fruits in some regions. Iron deficiency is difficult to control in orchards where soils have high alkalinity. Granulated fertilizers in modern close-planted commercial orchards are usually broadcast by machine a month or two before growth starts. Additional nitrogen sometimes is applied in heavy crop years to apple, pear, and citrus.

Crop Enhancement

Pollination

The stimulus of pollination, fertilization, and seed formation is needed to get good size, shape, and flavour of most of the fruits. (Banana, pineapple, and some citrus and fig varieties are exceptions.) Transfer of pollen from the anthers (male) to the stigmas (female) is accomplished in nature either by insects or by movement in air. It is common practice to bring beehives into the orchard during bloom. Rainy cold weather during bloom with little or no sunshine can deter activity of the honey bee (the key insect pollinator) and reduce fruit set appreciably. This is one of the main problems not fully solved by fruit researchers. Hand-pollination by daubing collected and preserved pollen onto the stigma (as is done with date palms) sometimes is practiced for other fruits, but this approach is not widespread.

Thinning

Removal of flowers or young fruit (thinning) is done to permit the remaining fruits to grow more rapidly and to prevent development of such a large crop that the plant is unable to flower and set a commercial crop the following year. Thinning is done by hand, mechanically, or chemically. With the date, the pistillate flower cluster is reduced in size at the time of hand-pollination. In the case of certain table grape varieties, some clusters are cut off. With the Thompson seedless grape, a combination of girdling the trunk bark and judicious application of gibberellin (growth regulating) sprays at blossoming gives excellent full bunches.

Young peach fruits are thinned by striking the branches with a padded pole or by shaking the entire tree for a few seconds with a well-padded motor-driven shaker arm grasping the trunk. Hand thinning of young apple and peach fruits once was also a common practice, but because of the expense and difficulty, there has been increasing use of chemical sprays as a substitute. Two kinds of sprays are used: (1) mildly caustic sprays applied during bloom, such as Elgetol in arid regions, or (2) sprays of growth-regulating substances such as 3-CPA (2,3-chlorophenoxy propionamide) applied within a few weeks after bloom in areas with late frosts.

Pest Control and Preservation

In many fruit enterprises, pest control is the most expensive and time-consuming growing practice. Where the concentration of fruit farms in an area warrants it, individual efforts are complemented by legislative measures including quarantine regulations to force removal of pest-laden, unattended orchards. Sometimes the most economical control procedure is biological in nature. There is increased research today to find and multiply parasites that kill fruit crop pests. Such biological methods are necessary as political pressures increase for banning DDT and other chemicals. Selection of varieties that are immune, resistant to attack, or tolerant to specific pests, is a biological control procedure also widely used. Chemical control procedures, however, are relied on most heavily. Air-blast spray or mist-application machinery covering 70 acres (28 hectares) of trees or more in a day is now in common use.

Harvesting and Packing

The proper time to remove a fruit from the tree or plant varies with each fruit and is governed by whether the product will be sold and consumed within hours, or stored for weeks, months, or even a year. Most fruits are harvested as close as possible to the time they are eaten. A few, of which banana and pear are outstanding examples, may be harvested while immature and still ripen satisfactorily. Orange, grapefruit, and some varieties of avocado may be "stored" on the tree for several months after they have attained good quality; this method cuts costs in handling and marketing.

Many fruits, including apple, pear, orange, lemon, and grapefruit, may drop from the tree during the last part of the maturation period. Preharvest drop of these fruits can be delayed by application of dilute sprays of growth-regulating substances like naphthaleneacetic acid (NAA). The chemical spray Alar [N-(dimethylamino) succinamic acid] applied four to six weeks after bloom on apple not only reduces fruit drop at harvest but increases red colour, firmness, and return bloom the next year, in addition to other advantages.

For the fresh market, most tree and bush fruits are still harvested by hand. For processing, drying, and occasionally for fresh market, mechanical motor-driven tree and bush shakers with appropriate catching belts, bins, pallets, and electric lifts reduce harvesting and handling labour. In years to come, machinery may make it possible to machine-harvest most fruits, with no more, and possibly less, damage than with hand picking.

The public has become increasingly particular about the appearance and quality of the product it buys. Hence, store managers and suppliers seek the best grades of fruits and nuts available, and growers make every effort to produce crops with attractive colour and smooth finish. Fruits are packed by government-controlled grades such as Fancy or Extra Fancy within given size limits and are so labelled on the carton or box, together with the source. Most fruits and nuts not meeting this standard of quality are processed or sent through channels using the lower grades and off sizes.

Small packages of plastic foam or wood pulp base holding four to six fruits covered and heat sealed with polyethylene plastic film are popular. These are delivered to stores in corrugated cartons holding a few dozen packages. Citrus, apples, and whole nuts or kernels also are packaged in polyethylene bags and delivered in cartons. Loose fruit may be sold in cell cartons and tray packs consisting of stacked form-fitting pulp trays in a "bushel size" box. Every effort is made to eliminate bruising.

Large truck-pulled containers with individually motor-driven refrigeration units, with or without controlled atmosphere (CO_2-O_2, to retard ripening), are loaded at the fruit source and trucked to their destination or are loaded on ships by derrick for overseas shipment. These sealed containers are also being used increasingly for bananas to reduce labour and handling and to deliver the product in better condition.

Air shipment of "vine- and tree-ripe" fruit (strawberries, figs, sweet cherries, pineapples, avocados) to distances as far as from California to Europe in a day or less is becoming increasingly common with the much larger and faster cargo planes and reduced air-freight prices.

Postharvest Physiology of Fruits

Fruit ripening is a form of senescence and signifies the final stage in fruit development. A fleshy fruit is the enlarged ovary of a flower (avocado) or additional floral parts such as in apple, pear, and pineapple. Usually fertilization, and sometimes pollination alone, stimulate the floral parts causing a rapid cell division that leads to differentiation and the formation of the fruit structure. During this stage fruits consist of small, young cells filled with protoplasm. When the young fruit has been stimulated, presumably by plant hormones that originate from the embryonic seeds, rapid cell expansion takes place. During this stage fruits gain rapidly in size and weight. The cells develop small cavities or spaces in their tissue (become vacuolated) and begin the process of foodstuff accumulation, which lends fruits their compositional diversity. Banana, apple, and date, for example, accumulate mainly carbohydrates. Avocado and olive store fatty materials. Important constituents of most fruits are organic acids such as malic acid, found in apple and pear; citric acid, found in citrus fruits and pineapple; and tartaric acid, found in grapes. Fruits are usually low in protein.

After cell expansion has slowed and become nominal, fruits enter the stage of maturity and undergo preparation for ripening. Some crops, such as pear and avocado, are harvested at the so-called mature-green state and allowed to ripen afterward. Most fruits are at a stage of incipient ripening before they are picked. Ripening is marked by rapid and dramatic changes that give fruits their attractive and edible character. Some of the familiar changes are softening, which results from degradation of cell wall substances; disappearance of a green background, because of chlorophyll degradation (as in pear, apple, and banana); appearance of coloured pigments such as the carotenoids—orange-yellow—and anthocyanins—red (as in orange, mango, and strawberry); a decrease in acidity and increase in the sugar content (orange, apple); and emission of the volatile substances that give many fruits their distinct aroma (as in banana, pear, and apple). In climacteric fruits (e.g., banana, pear, apple), ripening is accompanied by increased respiration. In nonclimacteric fruit (e.g., strawberry, cherry) this phenomenon does not occur.

It is thought that the transition from the mature to the ripe stage is brought about by certain "ripening" enzymes. Protein molecules act as catalysts. The activity of these enzymes leads first to various ripening reactions, and then to gradual deterioration of the fruit tissue.

Because ripening leads to tissue breakdown, fruits are considered a highly perishable commodity. Different fruits have varying degrees of postharvest longevity. While strawberries last only a week to 10 days, for example, apples or lemons can be stored successfully for as long as several months.

Postharvest life of fruits can be extended by refrigeration with or without a modified oxygen–carbon dioxide atmosphere. Most temperate-zone fruits can be held safely at 32 to 41 °F (0 to 5 °C), but many subtropical and tropical fruits, including lemon, avocado, banana, and mango, show signs of injury from being chilled in prolonged cold storage and consequently fail to ripen properly. Bananas do not tolerate temperatures below 53 °F (12 °C), while several avocado varieties can be stored at temperatures as low as 46 °F (8 °C).

Fruit life can be extended further by both refrigeration and controlled atmosphere (CA) storage in which oxygen is kept at about 5 percent and carbon dioxide at 1 to 3 percent, while temperature is held at a level best suited to the particular fruit. So-called CA storage is common today for apples and pears and is being adapted to other fruits. Controlled atmosphere and refrigeration in conjunction with the removal of ethylene gas (which emanates from fruits and speeds ripening) helps slow the ripening process considerably. Golden Delicious apples and some pears are shipped in polyethylene containers in which a desirable, modified atmosphere is created by the respiring fruit.

Drying is a standard practice for stabilizing the market movement of dates, figs, raisin grapes, prunes, and apricots. Canning is of paramount importance to the pineapple, peach, and pear industries (these fruits can be dried as well), and freezing is a means of stabilizing some of the most perishable fruits, including strawberry, raspberry, and blueberry.

Nuts are susceptible to mold, souring, staleness, discoloration, and rancidity. Cured and dried nuts are kept in prolonged cold storage under controlled temperature and humidity levels. Nuts also are stored and sold in vacuum packs of carbon dioxide-enriched atmosphere.

Waste Materials and other Uses

Apple wood is excellent for fireplace use, and cherry and certain other fruit woods are used for the finest household furniture. The dried residue from processing apples and citrus is made into feed for conditioning livestock for market, as are waste materials from many processed fruits. Apple pomace (waste material) is spread on the orchard floor with a manure spreader to help in soil conditioning and as a source of minerals.

Nutshells have many uses. Filbert shells are made into plywood, artificial wood, and linoleum; a mixture of shells with powdered coal and lignite makes cinder blocks; shells are used in making poisonous gases and gas masks, and as fuel and mulch. Cashew shell liquid, a skin irritant, is made into resins for varnishes; kills mosquito larvae; can be impregnated in wood as a varnish to preserve against insect attack; is used in automotive brake linings and clutch facings; is used as a laminating agent for paper, cloth, and glass fibres; and is used to treat cement floors and synthetic rubber to retard deterioration. Finely ground black-walnut-shell flour is used in plastic molding powder; as a glue extender; to prevent overheating of drills; to "sand"-blast jet engines; for polishing, burnishing, and deburring metal parts; for cleaning foundry molds; and to spray on tires for better traction. Pecan shells are used in place of gravel in cement walks and driveways; as fuel; as mulch and as a soil conditioner; in livestock bedding; as filler for fertilizers, feeds, etc.; in the manufacture of tanning agents, with charcoal and abrasives in hand soap; as a filler in plastic and veneer wood; and many of the same uses as black walnut shells. Some nutshells are made into beads, marbles, buttons, carving tools, ink, and ornament. The India clearing nut is cut open and rubbed on the inside of earthenware that will contain drinking water; the juice coagulates the water impurities which sink to the bottom. The nuts of the betel palm in the Far East and of the kola tree in West Africa are chewed for their stimulatory effects.

Use of Plant Growth Regulators in Fruit Production

Growth mainly refers to the quantitative increase in plant body such as increase in length of the stem and root, the no. of leaves, the fresh weight and dry weight etc. On the other hand, germination of seed, formation of flowers, fruits and seeds, emergence of lateral buds, falling of leaves and fruits are qualitative changes, referred to as development.

Growth and development of the plant body are controlled by two sets of internal factors, namely, nutritional and hormonal. Nutritional factors supply the plant necessary mineral ions and organic substances such as proteins, carbohydrates and others. These constitute the raw materials required for growth. However, utilization of these substances for proper development of the plant is controlled by certain chemical messengers, called plant growth substances or plant growth regulators, which in minute amounts increase or decrease or modifies the physiological processes in plants.

The term plant growth regulators' is relatively new in use. In earlier literature these were mentioned as Hormones. Hormone is a Greek word derived from hormao which means to stimulate. Now the term phytohormone is used in place of plant hormone.

Plant growth regulators or plant regulators are the organic compounds other than nutrients which modify or regulate physiological processes in an appreciable measure in the plants when used in small concentrations. They are readily absorbed and these chemicals move rapidly through the tissues when applied to different parts of the plant.

Plant hormones or phytohormones are also regulators but produced by the plants in low concentrations and these hormones move from the sit of production to the site of action. Therefore, the difference between the plant regulator and plant hormone is in that the former one is synthetic and the latter one is natural from the plant source.

The various types of growth regulating substances are:

- Auxins

- Gibberellins

- Cytokinins

- Ethylene

- Abscisic acid.

Auxins, Gibberellins and cytokinins are Growth promoters and Ethylene and Abscisic acid are growth inhibitors. Growth Retardants - These are chemicals which have common physiological effect of reducing stem growth by inhibiting cell division.

Growth regulating substances have many practical applications in horticulture and some of the most important uses are:

1. Propagation of plants: The most common use of plant regulators in horticulture is to induce rooting in stem cuttings and in air and soil layers.

2. Rooting of cuttings: Certain kind of plants may not successfully root under normal condition and with the aid of plant regulators; they can be easily made to induce rooting. The most commonly employed growth regulators for rooting are auxins like IBA, IAA, IPA and NAA. Among these chemicals IBA is most ideally used since, it is the most effective one.

Concentrations ranging from100-500ppm are used for long dip method of treatment of cuttings for 12-24 hours and high concentrations of 10,000 to 20,000 for quick dip method for a few seconds. The concentrations differ according to the type of cutting i.e. herbaceous, Semi-hard wood and hard wood cuttings. Applications in the form of dust as talcum preparation or in the form of a paste in lanolin are also used.

3. Layering: Another usage of plant regulators in plant propagation is in aiding rooting of air layering. Layering is the practice of inducing rooting on shoots/stems while it is still attached to the parent plant. This is practiced in fruit trees like guava, pomegranate etc. The main principle of layering is that a part of the aerial portion of the intact plant is girdled. This results in severing of phloem. Consequently, hormones and food substances coming from the leaves accumulate above the girdled portion. When the ring of bark is removed from the stem, the growth regulators like IBA or IAA in power or in Lanolin paste is applied at the distal end of the bark-removed portion to promote root formation.

4. Grafting and Budding: Grafting of plants is a widely used horticultural practice of multiplying the desired genotypes in mango, citrus and others. For this, a portion of the plant is inserted in to another plant of the same species or some times compatible plants of different species or genera. There are mainly two types of grafting: bud grafting and scion grafting. Whatever may be the method employed, the principle remains the same. When the cambium of a stock plant comes into physical contact with the cambium of a scion both from new xylem and phloem simultaneously together. Consequently, these become united and grow as one plant. Since, auxins have the property of promoting cell division of cambium these are often employed. Before grafting, either stock

or scion or both are dipped in auxin solution. This promotes an early union and consequently, a better success of grafted plants.

5. Control of flowering: The plant growth regulators are used for the regulation of flowering in certain crops. In pineapple flowering is irregular and harvesting becomes a problem and hence to regulate flower production, plant regulators are used. The treatment generally consists of pouring a required quantity of (50ml), the solution containing 0.25 to 0.5 mg of the chemical of NAA in the central core of plants. In recent studies, Cycocel and Alar at 5000ppm and Ethrel at 100-200ppm have been shown to induce flowering in mango during an off year. In Jasminum grandiflorum, the flowering period is extended by the application of Cycocel at 500ppm. Flowering can also be induced in certain vegetables such as radish, beet root and carrot with the application of GA.

6. Fruit set: Various growth regulators like IAA, IBA, IPA, NAA, 2, 4-D, 2, 4, 5-T and GA have been found to improve fruit set in many crops. Among these chemicals 2, 4-D and NAA (Planofix) have been found in general to be most effective in increasing the fruit set. The optimum concentrations for this purpose are 10-20 ppm of auxins and 10- 100ppm of GA in different crops. Spraying the flower cluster thoroughly 4-6 days after full bloom with 100 ppm GA increased the fruit set in grape. It has been found that in chillies spraying of Planofix @ 1ml in 4.5 litres of water at 60th and 90th day after planting is beneficial for good fruit setting.

7. Fruit drop: Losses resulting from pre-harvest drop of fruits have long been a serious problem. When the growth regulators have been put in to use in apples and pears, preharvest fruit drop can be checked by the application of 2,4-D and 2,4,5-T effectively. Pre harvest fruit drop in citrus is controlled with 2,4-D at a concentration of 20ppm2,4-D, 10- 15ppm of NAA and 2,4,5-T at 15 to 30ppm at pea stage and marble stage and 2,4D at 20ppm and 2,4,5-T at 10-15ppm in mandarins. At 10ppm and NAA at 20ppm have effectively prevented fruit drop in mango. Application of planofix containing NAA at pea seed —and marble size of the fruits completely controlled early fruit drop in Guava.

8. Parthenocarpy: Partenocarpic fruit set could be induced in a no. of vegetables like cucurbits, bhendi, brinjal, chillies and tomato and fruits like guava, straw berry, citrus, watermelon etc. IAA,IBA,NAA,NOA,NAD,2,4-D,IPA and GA are effective in different plants. Application of GA at 100 ppm induced complete seedlessness in grape varieties Viz., Anab-e-shahi,Pachadraksha etc. The problem of development of seeds in Poovan variety of Banana in Trichy area of Tamilnadu is controlled by application of 2,4-D at 25ppm in the bunches when the last hand is opened.

9. Fruit ripening: The plant growth regulators can be employed to hasten or delay fruit ripening. Plant growth regulators like 2, 4, 5-T at concentrations of 25 to 100ppm has been found to hasten the ripening in some varieties of plums and peaches. In banana ethrel treatment at 2500ppm induces ripening in 24 hours.

Application of 2, 4-D at 16ppm delays ripening in Washington navel oranges. In Calymirna fig maturity and ripening of the fruit is greatly hastened by spraying 2, 4, 5-T, while in apples in addition to this B-Nine also hastens ripening by about 1-4 weeks. Ethephon has been shown to hasten ripening in grapes.

In tomatoes all fruits on a plant won't mature and ripen at a time. This is a serious disadvantage for mechanical harvesting .Ethephon applied 1-2 weeks before harvest promotes degreening and ripening of tomatoes. Application of smoke is commercially employed to hasten and ripen bananas, the active ingredient responsible being ethylene. Ethyphon is also employed for degreeing and colour development of harvested fruits.

10. Fruit size and quality: Increase in berry size in Anab-e-shahi, Kismis and Bhokri varieties was reported when GA was applied at 40ppm at bud and flower stages. Higher concentrations resulted in the increase in the length of berries.

11. Sex expression: Plant regulators can be employed to modify the sex expression in crops. In cucurbitaceous vegetables the production of male flowers will be always more in number than the female flowers and this sex ratio can be narrowed down by the application of ethrel at 100 to 250ppm, if sprayed four times at weekly intervals commencing from 10 to 15 days after sowing. This growth regulator not only increases the number of female flowers to male flowers, but also produces female flowers at earlier nodes. Application of GA, the sex ratio is shifted towards maleness in several cucurbits.

Certain plant regulators are employed to induce male sterility in crop plants, so that such male sterile plants can be used as a female plant in the hybridization work. This process dispenses the expensive work. Complete male sterility in bhendi can be obtained by spraying with 0.4% of MH. A single spray one week before floral bud initiation offers male sterility for 10 days and a subsequent spray at floral initiation extends the effect to 22 days.

Preparation of Growth Regulators

1. Solution form: To prepare an alcoholic solution of any plant growth regulator, dissolve 1 gm. of growth regulator in 50 ml of ethyle alcohol or methyl alcohol or methylated spirit and then dilute this with an equal volume of water to make 100ml of solution containing 10,000 ppm of growth regulator.

This acts as stock solution for further dilutions with distilled or de-ionized water. Stored in well-stoppered bottles in a refregirator, the solutions retain their activity indefinitely.

2. Dust form: To prepare a dust containing 10,000ppm of growth regulator dissolve 1 gm of the regulator in 40 ml of methylated spirit of 95% alcohol and stir this into 100g of pharmaceutical talc to form a smooth paste. This should be done in a dark room away from strong light. Stir the paste while it is drying until it becomes a fine dry powder. This prepared dust remains active for six months or more if stored in a closed opaque container in a refregirator. From this stock, before using, dilute the growth regulator by mixing the stock with talc powder.

3. Lanolin pastes: These are particularly convenient for use in air layering but now regarded as an obsolete treatment for cuttings and are made by stirring the growth regulators into the molten lanolin and then allowing it to cool. To make a paste containing 5,00ppm of growth regulator melt 200gm of lanolin and thoroughly stir into this molten lanolin 1gm of required growth regulator. This prepared paste will keep indefinitely if stored in a well stoppered opaque glass vessel in a refrigerator.

Diseases of Fruits

Diseases of Mango and their Management

Anthracnose: Colletotrichum gloeosporioides.

Anthracnose symptoms occur on leaves, twigs, petioles, flower clusters (panicles), and fruits. The incidence of this disease can reach almost 100% in fruit produced under wet or very humid conditions. On leaves, lesions start as small, angular, brown to black spots and later enlarge to form extensive dead areas. Panicles develop small black or dark-brown spots, which can enlarge, coalesce, and kill the flowers. Petioles, twigs, and stems are also susceptible and develop the typical black, expanding lesions. On the lesions and dead portions, minute pink cushion shaped fructifications called acervuli are seen under moist conditions. Fruits may also drop from trees prematurely due to rotting. On immature fruits infections penetrate the cuticle, but remain quiescent until ripening of the fruits begins. Green fruit infections that take place at mature stage remain latent and invisible until ripening and carry the fungus into storage.

Favorable conditions: Under favourable climatic conditions of high humidity, frequent rains and a temperature of 24 - 32 °C coinciding with flowering favours anthracnose infections in the field. The pathogen survives between seasons on infected and defoliated branch terminals and mature leaves. Field infection in developing fruit leads to Quisecent infection/Latent infection. Later once the ripening starts, lesions begin to develop under post harvest conditions which affects fruit quality and leads to enormous loss.

Management

- Proper sanitation of orchard by periodical removal of fallen plant debris and pruning of trees eradicates the fungus and checks further spread of the disease.

- Maintaining tree vigour with proper irrigation and fertilization.

- Fungicide sprays should begin when panicles first appear and continue at the recommended intervals until fruits are picked.

- Spraying the trees twice with Carbendazim (0.1%) or Mancozeb (0.2 %) or combination of Carbendazim 12 % + Mancozeb 63 % @ 0.1 % at 15 days interval during flowering to control blossom infection and twice during pea nut stage to prevent fruit infection. Alternate sprayings of Carbendazim and Mancozeb to avoid development of resistance in pathogen to fungicides.

- Spraying five times with Pseudomonas fluorescens FP 7.

- (0.5%) from flowering until harvest at 3 weeks interval reduces anthracnose incidence and improves fruit quality.

- For post harvest anthrcanose, fruits are dipped in hot water at 50 C for 30 min. in combination with 0.05 % carbendazim.

Die Back/Fruit Stem End Rot - Lasiodiplodia Theobromae

The disease is characterized by drying of twigs and branches followed by complete defoliation, which gives the tree an appearance of scorching by fire. Tip die back disease occurs on the branches, trunk of infested trees that start drying slowly at first and suddenly branches become completely dried /killed resulting gummy substance oozes out or remains hanging on the tree. The dark area advances and young green twigs start withering first at the base and then the twig or branch dies, shrivels and falls called die back. This may be accompanied by exudation of gum. In old branches, brown streaking of vascular tissue is seen on splitting it longitudinally.

Stem end rot appears as rotting from pedicel end of fruit during riening until pathogen remains latent forming appressoria that remains quiescent in the subcuticular layer in green fruits (Quisecent infection/Latent infection).

Favorable conditions: High summer temperatures predispose the plant to attack by the pathogen. Relative humidity of 80% with temperature of 25.9 to 31.5 °C favor disease development. Survives in dead/diseased twigs, bark of the trees and fallen fruits.

Management

- Pruning of infected plant parts from 7- 10 cm below the infection site and pasting the cut ends with Bordeaux paste.

- Spraying the trees twice at 15 days interval with Carbendazim (0.1%) or combination of Carbendazim 12 % + Mancozeb 63 % @ 0.1 % during pea nut stage to prevent fruit infection.

- Fruits should be harvested with stalk (5 cm), otherwise, the opening must be sealed with wax.

Powdery Mildew - Oidium Mangiferae

1. The disease affects inflorescence, leaves and young fruits Symptoms: Appearance of a whitish, powdery growth of the fungus on leaves, panicles and young fruit which later turns brown and fall. The white growth can also be seen on the undersurface of young infected leaves which becomes distorted. Severe infection of young leaves results in premature leaf drop. On mature leaves, the spots turn purplish brown, as the white fungal mass eventually disappears. On developing inflorescence powdery growth leads to drying of flowers. Young fruits at peanut stage are covered with mildew that leads to corky tissue and drops.

2. Favorable conditions: Spread of the disease occurs when rains or mists accompanied by cooler nights during flowering especially when the weather is cool and dry. Minimum temperature of 13- 15 °C, maximum temperature of 23-25 °C with moderate relative humidity (64-72%) favours disease development. Pathogen survives in affected plant debris and under favourable conditions, air borne conidia is dessiminated by wind and attacks new flushes.

Management

- Pruning of diseased leaves and panicles.

- hree sprays of fungicides at different stages starting with Wettable Sulphur (0.2%) at the time of panicle initiation followed by Dinocap (0.1%) subsequently followed Tridemorph (0.1%) at 15-20 days interval.

- Sparying with mycobutanil @0.1% or Triademefon @0.1% or carbendazim @0.1% or Thiopahante methyl 0.1% found effective against disease.

Grey Leaf Blight - Pestalotia Mangiferae

Symptoms

Brown spots develop on the margin and at the tip of the leaf lamina which coalesce covering the leaf margin and becomes dark brown. Black dots appear on the spots which are acervuli of the fungus. If infection starts from tip, it advances on either side of mid rib and within 3-4 months severe defoliation results.

Favorable conditions: Heavy infection is noticed during the monsoon when the temperature is 20-25 °C and high humidity Conidia survive on mango leaves for over a year. Spreads through wind borne conidia and rain splashes. Wound leads to more disease.

Management

Removal of infected plant parts. Spraying one time with Copper oxychloride @ 0.25 % or Mancozeb @ 0.25% or Bordeaux mixture @ 1.0% at the visual appearance of disease.

Sooty Mould (Capnodium Mangiferae C. Ramosum)

The disease is common in the orchards where mealy bug, scale insects and hoppers are not controlled efficiently. Honey dew secretion by insects make the fungi produce mycelium which is superficial and dark and forms black encrustation on leaves. In severe cases, the trees turn completely black due to the presence of mould over the entire surface of twigs and leaves. The severity of infection depends on the honey dew secretion of the above insects. Presence of a black sooty mould on the leaf surface adversely affects the photosynthetic activity of the leaf and thereby fruit set is reduced.

Favorable conditions: Reduced ventilation favors sooty growth on leaves. High humidity spreads the disease within orchard. Honey dew secretions from insects stick to the leaf surface and provide necessary medium for fungal growth. Conidia spread by rain splashes. During rain sooty growth is washed away.

Management

Pruning of affected branches and their prompt destruction .Spraying systemic insecticides like to control insects. Spraying of 5 per cent starch (1kg Starch/Maida in 5 litres of water. Boiled and dilute to 20 liters) helps to control the disease as dried starch flakes removes the fungus.

Mango Malformation: Fusarium Moniliforme var. Subglutinans Symptoms

Three types of symptoms: Bunchy top phase, floral malformation and vegetative malformation. In bunchy top phase in nursery at 40-5 months old bunching of thickened small shoots, bearing small rudimentally leaves. Shoots remain short and stunted giving a bunchy top appearance. In vegetative malformation induces excessive vegetative branches of limited growth in seedlings. They are swollen with short internodes forming bunches of various size and the top of the seedlings shows bunchy top appearance. In malformation of inflorescens, shows variation in the panicle. Reduction in lenghth of primary axis and secondary branches of panicle makes the flowers to appear in clusters. Secondary branches are transformed into number of small leaves giving a witches broome appearance. Malformed head dries up in black mass and persist for long time. Such panicles dot not bear. The infection is localized.

Favorable conditions: Diseased propagated material spreads disease. The fungus does not sporulate insitu but sporulates on dried malformed panicles. The disease is severe in north west region at temperatures between 10-15 °C during December to January before flowering. Disease is mild in areas with 15-20 °C, sporadic bwteen 20-25 °C. Occurrence of malformation differs based on age of planrts. Plants at 4- 8 years are susceptible. In some cases mites have been reported to be carrying the fungus and cause spread.

Management

- Diseased plants should be destroyed.

- Use of disease free planting material.

- Incidence reduced by spraying 100-200ppm NAA during October.

- Pruning of diseased parts along the basal 15-20 cm apparently healthy portions followed by the spraying of Carbendazim (0.1%) or Captafol (0.2%).

Bacterial leaf Black Spot/Canker (Xanthomonas Campestris pv. Mangiferae-indicae)

Symptoms

This disease that affects all the above ground parts of plant, i. e., leaves, petioles, twigs, branches and fruits. The disease causes fruit drop (10-70%), yield loss (10-85%) and storage rot (5-100%). Many commercial cultivars of mango are susceptible to this disease. Bacterial leaf spot is noticed on the leaves as angular water soaked spots that become necrotic and dark brown and viscous bacterial exudates deposit on these necrotic portions that become corky and hard after drying. Sometimes, longitudinal cracks also develop on the petioles. Cankerous lesions appear on petioles, twigs and young fruits. The water soaked lesions also develop on fruits which later turn dark brown to black. They often burst open, releasing a highly contagious gummy ooze containing bacterial cells. The fresh lesions on branches and twigs are water soaked which later become raised and dark brown in colour with longitudinal cracks but without any ooze.

Favorable conditions: The bacteria enters through natural openings such as stomata, wax and oil glands, leaf and fruit abrasions, leaf scars, and at the apex of branches in the panicle. Periods of high humidity, surface wetness and wind accompanied with rain cause most rapid and maximum dissemination of bacteria. Survives in infected plant parts and spread through rain splashes and wind. Disease is rapid during rainy days.

Management: Field sanitation and removal of affected plant parts. Three sprays of Streptocycline (200ppm) or Agrimycin-100 (100 ppm) after first visual symptom at 10 days intervals. Monthly sprayings of Copper oxychloride (0.3%) checks the further spread. Removal of diseased fruits under storage. Dipping fruits in 200 ppm Agrimycin is effective.

Red Rust (Cepbaleuros virescens)

The disease is caused by an algae that causes reduction in photosynthetic activity. Initially the spots are greenish grey and velvety in texture which finally turn to rusty spots on leaves and twigs. Post sare initially circular, elevated and later coalese. Numerous filaments which may be sterile or fertile project outwards through cuticle. In severe cases, defoliation of leaves there by lowering vitality of the host plant reddish brown. After shedding the spore the algal matrix remains attached to leaf surface, leaving a creamy white mark at the original rust spot.

Favorable conditions: Disease is common in closed plantations. High humidity favours development of fruiting bodies.

Management: Supply of balanced nutrients to the plants and two sprays of Bordeaux mixture (1%) or Copper oxychloride (0.3%) in the month of July at 15 days interval Algal growth on leaves.

Integrated Disease Management Strategies

- Proper sanitation of orchards by removal of fallen plant debris and pruning of trees during the months of July –August eradicates the fungus and checks the spread of diseases.

- Application of Bordeaux paste on the cut ends of branches prevents entry of pathogens.

- During flowering, spray the trees twice with Carbendazim (0.1%) at 15 days interval to control blossom infection and twice during pea nut stage to prevent fruit infection by anthracnose and stem end rot. Alternate sprayings of Carbendazim and Mancozeb to avoid development of resistance in pathogen to fungicides. SAAF (Carbendazim 12 % + Mancozeb 63 %) @ 0.1 % can be recommended as an alternate for better control. Three sprays with Wettable sulphur (0.2 %) at 15 days interval starting from panicle initiation to prevent powdery mildew (sprays can be reduced depending on disease incidence).

- From fruit set until 15 days before harvest, alternate Carbendazim (0.1%) with a copper oxy chloride (0.3 %) every 14–28 days which takes care of fruit infections by anthracnose, stem end rot and bacteria.

- Spraying five times with Pseudomonas fluorescens FP 7 (0.5%) from flowering until harvest at 3 weeks interval reduces anthracnose incidence and improves fruit quality.

- Spraying of 5 per cent starch solution to remove sooty mould growth.

- Avoid injuries/damage on the fruits during picking and transit.

- Post harvest infection due to fungal pathogens are controlled by dipping the fruit in hot water (52 ± 10 °C) or hot water in combination with Carbendazim (0.05%) for 5 minutes.

Diseases of Banana and their Management

Fusarium wilt: Fusarium oxysporum Fsp. cubense.

Symptoms: Symptoms are characterized by yellowing of leaves from margins inwards towards mid rib. Petiole buckling takes place; as a result leaves hang around the pseudostem forming yellow skirt like pattern. The emerging heart leaf dies and only few erect leaves are seen on diseased plant. As the disease progresses, the younger leaves collapse until the entire canopy consists of dead or dying leaves. Longitudinal splitting of pseudostem is seen in advanced stages. The infected rhizomes when cut shows pinkish discoloration of vascular strands in pseudostem and brown discoloration of corm. The cut end of corm smells like rotten fish. Susceptible varieties are Rasthali, Karpooravalli, Neypoovan, Monthan.

Favorable conditions Primary inoculums is from infected suckers and pathogens in soil in the form of chlamydospores. Movement of infected plnating material, contaminated trash helps in long distance dispersal. Within field, irrigation water and farm implements are the major cause for spread of disease. Fungus gains entry through roots. Nematodes predispose the disease.

Management

- Considering the long-term survival of F. oxysporum f. sp. cubense in soil, proper field sanitation, use of disease free planting material/ tissue culture plants can reduce disease incidence.

- Severely affected plants should be uprooted and burnt. Highly infected soil should not be replanted with banana at least for 3-4 years.

- An integrated approach for management of Fusarium wilt has to be followed by use of pathogen free suckers, paiRing of suckers (Sucker treatment with Carbofuran granules@ 40 g/sucker before planting), Sucker dipping with 0.2 % carbendazim for 45 minutes at the time of planting , corm injection with 2% Carbendazim (3ml/plant) or capsule application with Carbendazim (50mg/capsule)/at 3rd, 5th and 7th month after planting will offer protection to the wilt susceptible cultivars.

- Application of Trichoderma viride or Pseudomonas fluorescens @ 50g/plant at the time of planting and at 4th, 6th and 8th month reduces the disease incidence.

Sigatoka Leaf Spot: Mycosphorella Musicola

Symptoms: Symptoms are characterized by oval to round necrotic lesions which first appear as pale yellow on lower surface of leaf. Correspondingly on the upper leaf surface, pale yellow specks appear which later on extends to form yellow oblong spots with or without yellow halo. When the disease progresses these spots further increase in size, join with each other forming large dead necrotic.

Areas on the leaf preventing photosynthetic functioning of the leaf. Destruction of mature and functional leaves in large number at the time of shooting leads to failure of bunches to fill out and ripen. Fruit set will be poor with reduced size, uneven ripening and angular shape having discoloured flesh. Sometimes premature ripening of banana bunches takes place in field itself.

Favorable conditions : The disease is influenced by in intermittent rainfall, high relative humidity and low temperature (23- 25 °C). Closer spacing, weeds, shade, frequent irrigation increase.

Management

- Proper, wider spacing must be practised.

- Severely infected plants and leaf blades should be removed and destroyed periodically.

- The orchard must be clean and free from weeds and grasses to avoid humidity build up.

- In the wet season, application of a protectant fungicide like Mancozeb @0.2% or chlorothalonil @0.1% every three to four weeks is recommended. During periods of high disease threat or extended wet weather, propiconazole, a systemic fungicide, can be substituted for Mancozeb. Foliar spray any one of the following fungicides commencing from October-November or mat months coinciding with rainfall at monthly interval. Carbendazim 1 g/lit., Benomyl 1 g/lit., Mancozeb 2 g/lit., Copper oxychloride 2.5 g/lit., Ziram 2 ml/lit, Chlorothalonil 2 g/lit. Alternation of fungicides prevents fungicidal resistance. spray Propiconazole 1 ml/lit or 0.5 ml/lit along with petroleum based mineral oil 10ml/lit or Pseudomonas flourescens (0.5%).

Rhizome Rot or Soft Rot or Tip Over Disease: Erwinia Caratovora Sub Sp. Caratovora

Symptoms: Rhizome Rot, also known as Soft Rot or Tip Over Disease is commonly observed during the first 3–4 months after planting under high temperature conditions, especially during late summer and rainy season. The heart leaves become shrivelled, brown and dried and the adjacent leaves are reduced in size. Rotting of pseudostem base and upper portion of corm accompanied with foul smell. Severely affected plants show breaking of pseudostem at base leaving infected and rotted corm in soil. Slimy fluid of bacteria oozes out from the cut portions of corm and pseudostem. Young plants of 1-3 months are more susceptible during summer months.

Favorable conditions: The bacteria survives in soil and infected plant residues and they enter the plant through wounds and spread across fields through water and infected planting material.

Management

- Management of Erwinia rot begins with the selection of suckers from diseases free field.

- Sucker dip in Copper oxy chloride (Blitox @ 5.0 g/l) + 300ppm streptomycin sulphate for 30 min or carbofuran @ 40 g/plant + Copper oxy chloride drenching or dipping of suckers in 5000 ppm of Copper oxy chloride for 30 min., reduced rhizome rot in banana.

- Intercropping with sunhemp in the interspaces till 4 MAP can further reduce disease by lowering the soil temperature and can add nitrogen source to soil.

- Under field conditions, soil drenching streptomycin sulphate (500 ppm), Streptocycline 500 ppm and Copper oxy chloride 2000 ppm reduce the rhizome rot disease incidence.

- Also dipping the suckers in Copper oxy chloride (40 g/10lit) Streptocycline (3g/10 lit.) for 30 min. before planting.

- Application of bleaching powder @6 g/plant followed by immediate irrigation at the onset of disease.

Moko Wilt: Ralstonia Solanacearum

Symptoms: Initial symptoms are characterized by yellowing of the inner lamina close to the petiole which is followed by wilting of inner leaves. The lower leaves become yellow which progresses upwards.Petiole break at this junction and droop around the pseudostem. If diseased suckers are planted, the terminal leaves become necrotic and plant dies. Young sword sucker from diseased plant show wilting and blackening. When pseudostem is cut pale yellow to dark brown vascular discoloration of strands are seen along with yellowish brown oozing of bacteria from cut ends. In some cases no external symtoms are produced until bunch development. The presence of yellow fingers in green bunches and inner pulp exhibits dark brown discolouration. Fruit rot and fruit stalk discoloration , wilted or blackened regowth of suckers, dead male flower buds and peduncles with vascular discoloration are other symptoms of Moko disease.

- Reduce disease by lowering the soil temperature and can add nitrogen source to soil.

- Under field conditions, soil drenching streptomycin sulphate (500 ppm), Streptocycline 500 ppm and Copper oxy chloride 2000 ppm reduce the rhizome rot disease incidence.

- Also dipping the suckers in Copper oxy chloride (40 g/10lit) + Streptocycline (3g/10 lit.) for 30 min. before planting.

- Application of bleaching powder @6 g/plant followed by immediate irrigation at the onset of disease.

Moko Wilt: Ralstonia Solanacearum

Symptoms: Initial symptoms are characterized by yellowing of the inner lamina close to the petiole which is followed by wilting of inner leaves. The lower leaves become yellow which progresses upwards. Petiole break at this junction and droop around the pseudostem. If diseased suckers are planted, the terminal leaves become necrotic and plant dies. Young sword sucker from diseased plant show wilting and blackening. When pseudostem is cut pale yellow to dark brown vascular discoloration of strands are seen along with yellowish brown oozing of bacteria from cut ends. In some cases no external symtoms are produced until bunch development. The presence of yellow fingers in green bunches and inner pulp exhibits dark brown discolouration. Fruit rot and fruit stalk discoloration, wilted or blackened regowth of suckers, dead male flower buds and peduncles with vascular discoloration are other symptoms of Moko disease.

Favorable Conditions: Pathogen is soil borne and survives in infected debris. Spread is by irrigation water, farm implements and infested soil and insects.

Management: Early detection and destruction of the suspected plants may help in preventing the spread of the disease. All the tools used for pruning and cutting should be disinfected with formaldehyde. As the insects can carry the disease causing bacterium on the male flowers, removal of the male flowers as soon as the last female hand emerge help in minimising the spread of the disease.

Viral Diseases of Banana

Banana Bunchy Top Disease: Primary transmission is by infected suckers and secondary transmission by the aphid vector Pentalonia nigronervosa.

Infected suckers putforth narrow leaves, which are chlorotic and exhibit mosaic symptoms. The affected leaves are brittle with their margins rolled upwards. Characteristic symptom of bunchy top virus is the presence of interrupted dark green streaks along the secondary veins of the lamina or the midrib of the petiole. The dark green, hook like extensions of the leaf lamina veins can also be seen between the midrib and the lamina. The diseased plants remain stunted and do not produce bunch of any commercial value. The suckers that develop after a 'mother' plant has been infected with BBTV are usually severely stunted, with leaves that do not expand normally and remain bunched at the top of the pseudostem. Therefore the disease can result in a 100% yield loss. When very late infection occurs in the season, the plant would show dark green streak on the tip of the bracts of male flower bud. The aphids are found clustered around the unfurled heart leaf and leaf sheaths. The aphids produce honeydew which attracts ants around the plant. The BBTV symptoms are visible only 22- 23 days after infected by the virus.

Infectious Chlorosis - Cucumber Mosaic Virus

Primary transmission is by infected suckers and secondary transmission by the aphids Aphis gossypii and Aphids maidis. The disease is characterized by typical mosaic symptoms on the leaves. Mosaic plants are easily recognised by their dwarf growth and mottled, distorted leaves. The earliest symptoms appear on young leaves ass light green or yellowish streaks and bands giving a mottled appearance. Necrosis of emerging cigar leaves leading to varying degree of necrosis in the unfurled leaf lamina. Internal tissues also show necrosis. Plants do not produce bunches but act as a virus reservoir. Affected plants throw small bunches with malformed fingers uneven riening has been associated with this virus.

Banana Bract Mosaic Virus (BBrMV)

Primary transmission is by infected suckers and secondary transmission by aphid Vectors Aphis gossypii, and Rhopalosiphum maidis. Purple coloured spindle shaped streaks on pseudostem. Clustering of leaves at crown with a travelers palm appearance, elongated peduncle and half filled hands. Bunches have unusually long peduncle, chocking of bunches, raisedcorky growth on peduncle is also observed. Bracts show thick purple longitudinal streaks. Dark reddish spindle shaped streaks are seen on pseudostem.

Banana Streak Virus (BSV): Primary transmission is by infected suckers and secondary transmission by the mealy bugs Planococcus citri and Saccharicoccus sacchari. A prominent symptom exhibited by BSV is yellow streaking of the leaves, which becomes progressively necrotic producing a black streaked appearance in older leaves. Necrotic streaks are observed on midrib, petiole and pseudostem. Bunch choking, abortion of bunch and seediness in fingers are observed. Delayed fruit bunch emergence, reduction in fruit size, malformation of fingers when floral initiation and early bunch development coincide with a period of increased virus synthesis. Sometimes necrosis of emerging leaves internal necrosis of the pseudostem and plant death.

Integrated Management for Virus Diseases

- Eradication of infected plants and sword suckers.

- Strict quarantine measures to be followed.

- Ensure virus free planting materials.

- Avoid cucurbitaceous crops around banana field.

- Spraying dimethoate or methyl demeton 0.1 % for vector control.

- Destroy bunchy top virus affected plants by capsule application of 200 mg of 2,4 -D/capsule in to corm (7 cm deep) using capsule applicator or inject 5 ml of 2,4-D solution (125 g/lit.) in pseudostem using injector.

- Control vector by spraying 1ml Methyl demeton (or) 2 ml Monocrotophos (or) 1 ml Phosphomidon in 1 litre of water (or) injection of Monocrotophos 1 ml/per plant (1 ml diluted in 4 ml water) at 45 days interval from 3rd month till flowering.

- Removal of weeds.

References

- What-is-pomology: wisegeek.com, Retrieved 5 March, 2019

- Fruit-Crop: thefreedictionary.com, Retrieved 14 June, 2019

- Classification-of-fruits: biocyclopedia.com, Retrieved 25 January, 2019

- Mango-farming-guide: agrifarmingtips.com, Retrieved 18 May, 2019

- Brodsky, Allen B (1978). CRC Handbook on Radiation Measurement and Protection. 1. West Palm Beach, FL: CRC Press. p. 620 Table A.3.7.12. ISBN 978-0-8493-3756-7

- Major-Producers-of-Banana: agrifarmingtips.com, Retrieved 9 February, 2019

- Smith, James P. (1977). Vascular Plant Families. Eureka, Calif.: Mad River Press. ISBN 978-0-916422-07-3

- Apple, fruits, crop-production, agriculture: vikaspedia.in, Retrieved 17 August, 2019

- Hyam, R. & Pankhurst, R.J. (1995). Plants and their names : a concise dictionary. Oxford: Oxford University Press. p. 329. ISBN 978-0-19-866189-4

- Orange-farming-guide: agrifarmingtips.com, Retrieved 7 April, 2019

- Fruit-farming: britannica.com, Retrieved 21 January, 2019

Chapter 4

Floriculture: A Comprehensive Study

Floriculture is a branch of horticulture that is concerned with the cultivation of ornamental and flowering plants for gardens and floral industry. It includes the cultivation of bedding plants, flowering pot plants and houseplants. Some of the major flower crops are rose, chrysanthemums, lilium and marigold. This chapter discusses in detail the processes and practices related to floriculture.

Floriculture is the study of growing and marketing flowers and foliage plants. Floriculture includes cultivation of flowering and ornamental plants for direct sale or for use as raw materials in cosmetic and perfume industry and in the pharmaceutical sector. It also includes production of planting materials through seeds, cuttings, budding and grafting. In simpler terms floriculture can be defined as the art and knowledge of growing flowers to perfection. The persons associated with this field are called floriculturists.

Worldwide more than 140 countries are involved in commercial Floriculture. The leading flower producing country in the world is Netherlands and Germany is the biggest importer of flowers. Countries involved in the import of flowers are Netherlands, Germany, France, Italy and Japan while those involved in export are Colombia, Israel, Spain and Kenya. USA and Japan continue to be the highest consumers.

Demand and Supply

The demand for flowers is seasonal as it is in most countries. The demand for flowers has two components: a steady component and a seasonal component. The factors which influence the demand are to some extent different for traditional and modern flowers.

(i) Traditional Flowers: The steady demand for traditional flowers comes from the use of flowers for religious purposes, decoration of homes and for making garlands and wreaths.

(ii) Modern Flowers: The bulk of the steady demand for modern flowers comes from institutions like hotels, guest houses and marriage gardens. The demand is concentrated in urban areas. With increasing modernization and globalization the demand for modern flowers from the individual consumers is likely to grow enormously as the trend of "say it with flowers" is increasing and the occasions which call for flower giving will continue to present themselves. Although there is an increasing demand for modern flowers from individuals, institutions continue to be the dominant buyers in the market. The price of these flowers also depends on their demand and varies accordingly.

Green House Technology for Flower Production

In present scenario of increasing demand for cut flowers protected cultivation in green houses is the best alternative for using land and other resources more efficiently. In protected environment suitable environmental conditions for optimum plant growth are provided which ultimately provide quality products. Green House is made up of glass or plastic film, which allows the solar radiations to pass through but traps the thermal radiations emitted by plants inside and thereby provide favourable climatic conditions for plant growth. It is also used for controlling temperature, humidity and light intensity inside. On the basis of basic material used, building cost and technology used, green houses can be of three types:

1. Low-cost greenhouse: The low-cost green house is made of polythene sheet of 700 gauge supported on bamboos with twines and nails. Its size depends on the purpose of its utilization and availability of space. The temperature within greenhouse increases by 6-10 °C more than outside.

2. Medium-cost greenhouse: With a slightly higher cost greenhouse can be framed with GI pipe of 15 mm bore. This greenhouse has a covering of UV -stabilized polythene of 800 gauge. The exhaust fans are used for ventilation which are thermostatically controlled. Cooling pad is used for humidifying the air entering the chamber. The greenhouse frame and glazing material have a life span of about 20 years and 2 years respectively.

3. Hi-tech greenhouse: In this type of green house the temperature, humidity and light are automatically controlled according to specific plant needs. These are indicated through sensor or signal-receiver. Sensor measures the variables, compare the measurement to a standard value and finally recommend to run the corresponding device. Temperature control system consists of temperature sensor heating/cooling mechanism and thermostat operated fan. Similarly, relative humidity is sensed through optical tagging devices. Boiler operation, irrigation and misting systems are operated under pressure sensing system. This modern structure is highly expensive, requiring qualified operators, maintenance, care and precautions. However, these provide best conditions for export quality cut flowers and are presently used by large number of export units.

Scope

There is a good scope for commercial floriculture. The important factors which decide the scope for Commercial Floriculture are soil, climate, labour, transport and market. Almost all big cities are developing very speedily to accommodate this fast growing population, cement concrete, jungle is also developing at the same rate and thus people are now realizing the importance of open space, parks and garden for relaxation, peace of mind, recreation and unpolluted air. Thus, to meet out all these problems bio-aesthetic planning is essential, which runs hand in hand with town planning. In modem life floriculture garden in the country yard is an integral part of the modern life and thus ornamental plants have found a place in home gardening.

As far as flower trade is concerned i.e. for cut flowers and loose flowers, it is growing very well in our state because these cut flowers are used for vase decoration and now-a-days there is a craze for indoor decoration. As far as loose flowers are concerned these are mainly used for preparation of gajara, veni, garland and bouquets and thus demand of flowers for these purposes is unending. Thus, taking into consideration the different points i.e. bio-aesthetic planning, floral garden, indoor decoration, social functions and religious functions the demand for floricultural plants is increasing day by day and to meet out the same there is a good scope for growing and raising of

Ornamental or Floricultural plants. When Flower Trade is concerned; different flowers like Rose, Chrysanthemum, Gladiolus, and Tuberose are demanded in the market as cut flowers. While Aster, Gaillardia, Marigold, Chrysanthemum, Jasmines, Tager Nerium as loose flowers.

Goals

1. Crop improvement - Genetic resource enhancement, evaluation and conservation.

2. Breeding of novel colour, short duration, temperature and drought tolerant ornamental cultivars.

3. Breeding of dwarf cultivars of high value flowers/foliage plants.

4. Strengthening/standardization/popularization of F1 hybrid seed production in important annual flowers.

5. Exportable 'made in India' varieties using molecular breeding and advanced techniques.

6. Production technology.

7. Development of agro-technologies for open field and protective cultivation – region specific, energy efficient, low cost and ecofriendly production systems.

8. Commercialization of 'Specialty Flowers' as new flowers to attract buyers and market demand.

9. Digitalized spray and fertigation schedules.

10. Production of quality planting materials through tissue culture.

11. Improved media, new cladding materials and alternate energy and light sources.

12. New generation molecules for enhancing blooming span, shelf life, etc.

13. Mechanization of planting to packaging and storage.

14. Post-harvest technology - Standardization of operation procedures for isolation of pigments, essential oils and natural colours/dyes.

15. New flower crops suitable for making dry flowers and technologies for efficient drying.

16. Natural dyes for pot- pourries and other flower arrangements.

17. Post-harvest engineering for improved packaging and low cost storage to reduce perishability.

18. Research on anti-senescence technology to delay senescence.

19. Landscaping - Improved grass species for turfs.

20. Urban greening technologies and vertical gardening.

21. New plants suitable for landscaping.

22. Plant health management.

23. Healthy landscape and beautiful flowers.

24. An inventory of pests and pathogens of flower crops.

25. Digitalized pest and disease forecasting expert system.

26. Prevention/management/eradication of the emerging pests and pathogens under changing climatic scenario.

27. Crop protection strategies with reduced pesticide inputs.

28. Multiple pest and disease resistant flowering plants.

29. Assured plant Biosecurity by generating database on potential pest and diseases of flower crops, diagnostics and containment techniques and information sharing.

30. Quality floriculture products meeting international sanitary and phytosanitary standards.

Flower Crops

Rose

Rose is the most popular flower the entire world over from times immemorial. Originally its fragrance attracted man's attention but later its colour, shape, size, recurrent flowering and vase life, etc. became the equally important attributes. To the naturally existing varieties in different agro-climatic localities, the rose enthusiasts added much more creations by utilizing the tool of hybridization for new and desired characters to incorporate maximum good qualities in a single variety. In the recent years another important pathway called genetic engineering is also being practiced to improve the qualities of available plant varieties and plant species. Although much success has been achieved in this direction so far efforts are continuously being done to develop still better varieties with the result that the number of varieties grown at any time indifferent agro-climatic situations is so enormous that any new rose enthusiast is usually puzzled to make a list of his choice. It is, therefore, convenient to know the major groups of rose cultivars commonly grown and which have persisted for several years in nature also.

Roses, thus, can be classified in different ways e.g. on the basis of purpose of growing viz. Garden roses, cut-flower roses, landscape roses, exhibition roses, fragrant roses, etc. on the basis of growth habits viz. Tall, medium, dwarf, climbers, ramblers, etc. or on the basis of botanical characteristics viz. Species roses. For practical purposes, however, certain criteria selected partly from growth habits and partly from botanical characteristics are the most suited for identifying the rose cultivars. Accordingly, the cultivated or popularly grown roses are usually and broadly grouped as follows:-

Large Flowered or Hybrid Tea (H.T.) Roses

Roses in this group are thin to medium thick bushes with height varying from about 75 to over 200 cm. Their name, Hybrid Tea Roses, is derived from the old naturally occurring roses giving fragrance resembling to that of a freshly opened tea chest and which were used in the hybridization programmes to develop the modern roses. The plants are usually vigorously growing, perpetually flowering with large (6-10 cm. wide) blooms having 15-20 to 70 –80 petals with attractive shapes and colours. Although most of them have no or very little fragrance a few emit strong scents characterised as fruity, spicy, sweet, etc. The flower stems are medium to long (30 to 100 cm) and in most cultivars every stem bears only a single blook. This is the most widely grown type of roses for home or public gardens, exhibition purposes and also for the cut flower trade. On the basis of plant growth habit, colour and shape of the bloom and fragrance etc. there are several hundred varieties or cultivars in this group suited to different agro-climatic conditions.

Clusters Flowered or Floribunda Roses

These are medium to very thick bushes – with height ranging from 50 to 120 cm. The blooms are of medium size having 6-8 to 20-25 petals, every stem often bearing 4 to 6 or even much more blooms. The stems are small to medium in length. Since abundant blooms are commonly produced in every flush, the plants exhibit excellent beauty of nature. Several good varieties with varying colour shades are available in this group also. These are most suitable for planting along the roads in beds and for landscape designing as well.

Miniature Roses

As the name suggests, these roses can be well described as miniature floribundas. The plant height is about 30 to 50 cm, with tiny delicate leaves and flowers. When in full bloom the plants get fully covered with small attractive colourful flowers in thick clusters. These roses also make excellent pathway borders and individual potted plants.

Climbing Roses

These are mostly naturally growing wild roses with thick vine-like growth habit. The blooms are of medium size and they are usually produced in clusters. The blooming period is also short and seasoned as compared to that of the other three groups described earlier.

Besides the four groups of roses described above, the wild roses like 'Polyanthas', 'Ramblers', etc. are also seen in some gardens but the blooming period being short and seasonal and due to shy blooming nature these roses are not very popular.

Although hundreds of excellent rose varieties have been developed so far and are being grown popularly, the rose hybridizes are continuously trying to evolve still better cultivars suitable for different purposes and different purposes and different agro-climatic situations and several new varieties are released every year in different parts of the world.

Another important point to be noted in this respect is that howsoever excellent a variety may be, it may not be so under all circumstances. Climate and Soil of a locality profoundly affects the performance of a variety.

Cultivation of Rose

The Main Steps for Growing Roses are:

Situation

The ideal position for planting roses is sunny, open and protected by hedge, fence or building against strong winds. It should be quite away from shade or from the underground roots of trees.

Soil

Roses require fertile and clay loam and loam soils. Soil should be deep having good water holding capacity with proper drainage. Roses do well in soils having pH between 6.0 to 7.5 but it can also grow satisfactorily in alkaline soil with pH up to 8.4. The soil pH can be brought in safe limits by adding gypsum or other acidifying agents in alkaline soil whereas pH of acidic soil can be raised by adding well ground dolomite lime stone.

Layout

Rose is a beautiful flower; hence, it should be displayed in an attractive manner. The beds can be made of formal designs or informal designs keeping the style of layout of the garden in view. Formal shapes of beds like square, rectangular, circular, U-shape or L-shape can be prepared. For informal designs, the beds of informal shapes suiting to the design should be prepared.

Planting

The place selected for planting roses should be dug thoroughly to a depth of 90-120 cm and kept open for few days. The soil should be dried and refilled with 10-15 kg/sq. m. well rotten farm yard manure and good garden soil at the top.

The spacing between plants varies with the vigour of the variety but generally H.T. varieties can be planted at the distance of 75 cm from each other while for the varieties of floribundas which are used for massing, a distance of 60 cm can be kept.

For planting roses, best time is from end of September to middle of October but it can be extended up to November. At the time of planting roses, the soil of the size of earth ball should be removed from the bed and plant should be placed in this pit.

Soil should be refilled and well pressed. Care should be taken that bud union is just above the ground. Light pruning i.e., tipping back of the branches should be done. After planting, frequent irrigation, removal of root suckers and manuring should be done.

Propagation

Roses are commonly propagated by 'T' or shield budding on the rootstock. The common root-stock used is Edouard rose (R. bourboniana) or R. multi-flora. Recently R. indica odorata has been found better than the former rootstocks.

Thorn-less rootstock is also getting popular. These rootstocks are easily propagated by hard wood and semi hardwood cuttings which are prepared in September-October at the time of annual pruning. When these cutting attain pencil thickness which takes about a year, they are ready for budding.

Roses are grouped into different classes according to the height at which budding is done on the rootstock. For Bush roses budding is done at 7.5-10 cm above the ground whereas for standard rose budding is done at the height of 100 cm.

These are more useful in the garden because flowers are produced 1 m above the ground and are suitable for both formal and informal layouts. Sometimes Half standard is also produced by budding at 50 cm. Weeping Standard roses are prepared by budding of climbing roses at 100 cm above the ground. In Weeping roses branches of climbing roses droop down and create artistic effect.

These rootstocks are easily budded by 'T' method and there is a high percentage (about 90-95%) of success. The best budding time is from November to February and after union of buds, budding starts growing. The upper portion of rootstock is cut into installments. By the time plants are ready for transplanting it completes about two years.

The new method of cuttage buddage is followed to reduce this long period. In this method, cuttings are budded immediately and planted in sand or burnt rice husk media for rooting under polythene cover in the month of December or January.

It takes about 3-4 weeks for cutting to strike roots and bud to grow. High success can be obtained by treating the cuttings and buds with 1000 ppm of IBA so that in one year plants are ready for planting. The miniature roses are propagated through cuttings.

Manuring

Many arbitrary recommendations are available about feeding of roses. Roses should be fed with both organic and inorganic sources. One hundred gram of mixture containing groundnut cake—5 kg, bone-meal—5 kg, ammophos (11: 48)—2 kg, Ammonium Sulphate—1 kg, super phosphate (single)—2 kg and Potassium Sulphate— 1 kg should be applied per bush for better results.

Results obtained at PAU reveal that addition of 60 g N, 20 g of P_2O_5 and K_2O should be applied per sq. m. containing nine plants. These fertilizers should be applied in two splits i.e., half amount of N, full dose of P and K at the time of prunning and remaining half one month after the first application. In the market many ready-made rose mixtures are being sold which can also be applied.

For better effects foliar feeding of roses has also been found very useful. Foliar sprays of 30 g of a mixture of 2 parts urea, 1 part Di-hydrogen ammonium phosphate, 1 part potassium nitrate and 1 part of potassium phosphate in 10 L of water at weekly or fortnightly interval, improve growth and flowering of roses. A weekly spray of 30 g urea/10 L of water, with an insecticide is also recommended.

Irrigation

Water requirement of roses depend upon soil type and seasons. Light soils require more frequent irrigation than heavy soils. During summer, water requirement is more than winter. Therefore, irrigation is adjusted in a way that soil is moist but not wet. During rainy season, watering is generally not done except during drought period. During winter, irrigation is done at about 7-10 days interval whereas during summer it should be done at an interval of 5-6 days.

Pruning

It is most important and technical aspect of rose growing. Root suckers should be removed whenever they appear. Every year rose plants produce some shoots which become weak by the young shoots growing out and become bushy in appearance. If not pruned properly, plants produce poor quality of flowers and become unproductive.

A judicious pruning is done which removes those plants which would become ultimately useless, rob and endanger the healthy wood. Pruning also helps in securing healthy and clean canes and production of quality blooms and availability of light and air to central parts, thus maintaining the health of plants.

The ideal time of rose pruning is when rains are completely over and winter is approaching. Most of the varieties take about 60-65 days of blooming after pruning. Therefore, to secure flowers for particular occasion, pruning can be adjusted accordingly.

Pruning should be done systematically by cutting out all dead wood and removing all weak and spindly growth. Overcrowding of shoots at the center should be thinned out to keep the center open. If too many shoots are left after thinning out process, reduce their number to few strong ones.

The canes are shortened according to the type of which it belongs. Pruning should always be done to an eye pointing outward and cut should be covered with fungicidal paint to check the infection of die-back pathogens.

Hybrid Teas and Teas are pruned to 5-6 eyes but for exhibition purpose severe pruning is done to 3-4 eyes. In case of Floribundas, dwarf polyanthus and miniatures, pruning consists of removal of all weak and diseased growth and also the ends of branches should be tipped back. In climbers, in-laced stems and all weak growth should be removed. Old stems may be cut after 3-4 years.

Chrysanthemum

Chrysanthemums are next to roses in popularity and have been in cultivation for more than 2,500 years. There are thousands of varieties now in cultivation in different countries and more than 3,000 varieties are being grown in English gardens alone.

Classification

Chrysanthemum varieties are classified into seven main groups, namely, the Incurved (like a perfect ball), Reflexed (with drooped florets), Incurving (in which the petals incurve loosely and irregularly), Anemone (having single petals and a tubular central disc), Pompon (with very small-sized flowers), Singles (having a central disc and five petals or ray florets), Miscellaneous like Spider (with a hook at the tip of petals), Spoon (with a spoon-like tip of petals), Koreans (having small single or double flowers with a visible central disc) and Rayonnantes (having quilled petals).

Varieties

There are numerous varieties of chrysanthemum, the flowers of which may be white, cream, yellow, red, terracotta, bronze, maroon, lilac, mauve, pink or purple. A few commonly grown varieties are Snowball (white, incurved), Sonar Bangla (cream, incurved), Rose Bowl (deep rose-pink, incurved), Alfred Simpson (crimson-red, bronze reverse incurved), Pink Cloud (pink, incurved), Ajina Purple (rose –purple, incurving), Bronze Turner (bronze, incurving), Coronation Pink (pale pink, reflexed), Duke of Kent (white, reflexed), Mahatma Gandhi (mauve, tubular, reflexed), Kasturba Gandhi (white, tubular, reflexed) and Valiant (large, white, reflexed). A few other important varieties include tubular, Kikubiori (large, yellow, incurved), Angel Bell (large, soft, pink, incurved), Shen Meigetsu (yellow, incurved), Grape Bowl (wine-red, incurving), Tokyo (white, spider), Rupasi Bangla (white, spider, dwarf), Flirt (deep maroon, double Korean type), Birbal Sahani (white, globular, pompom), Mountaineer (large, yellow, incurved), and William Turner (white, incurved).

- Anemone: These daisy-like blooms feature long, tubular florets clustered around a tight button center. They form a 4-inch bloom in single or multiple colors. Popular varieties include - Dorothy Mechum, Purple Light and Angel.

- Decorative: Florists use decorative class chrysanthemums in floral arrangements. The 5-inch plus blooms have a flat appearance as the florets gradually get longer from the center out. Popular varieties include - Fireflash, Coral Charm and Honeyglow.

- Irregular Incurve: Incurve blooms feature florets curving inwards. Irregular incurve chrysanthemums feature large blooms between 6 to 8 inches. The florets curve in and cover the center of the flower. A few florets at the bottom of the bloom add fringe to the stem. Popular varieties include - Luxor, Blushing Bride and River City.

- Intermediate Incurve: The florets of an intermediate incurve mum don't cover the center of the bloom. With shorter florets curving inwards, the less-compact bloom of an intermediate incurve only reaches a maximum 6 inches. Popular varieties include - Apricot Alexis, Candid and Pat Lawson.

- Regular Incurve: Regular incurve chrysanthemum blossoms are tight, smooth globes of inwardly curving florets. Each bloom is between 4 to 6 inches in diameter. Popular varieties include - Gillette, Moira and Heather James.

- Pompom: Resembling the regular incurve, Pompom chrysanthemums are only 1 to 4 inches. The tight blooms are common in floral arrangements. Popular varieties include - Rocky, Yoko Ono and Lavender Pixie.

- Quilled: Show-stopping quilled chrysanthemums feature long, tubular florets that open to a spoon shape or slight downward curve at the end. Their spiky appearance often mimics other types of chrysanthemums. Popular varieties include - Seatons Toffee, Mammoth Yellow Quill and Muted Sunshine.

- Single and Semi-Double: These daisy look-a-likes feature one or two rounds of ray florets around a compact center. Their total plant size is between 1 to 3 feet, making them ideal for small spaces and borders. Popular varieties include - Rage, Icy Island and Crimson Glory.

- Spider: Spider chrysanthemums are well known for their long, spiky florets of single or multiple colors. The tubular florets resemble spider legs and can go in all directions. The delicate and exotic appearance creates a focal bloom in your garden. Popular varieties include - Evening Glow, Symphony and Western Voodoo.

- Spoon: Spoon chrysanthemums have a button center surrounded by ray florets featuring a spoon shape at each tip. They are often mistaken for single chrysanthemums , but the difference lays in the slight curve. Popular varieties include: Kimie, Fantasy and Redwing.

- Reflex: The bloom of a reflex mum is slightly flat with florets that curve downward. The crossing of the florets produces an interesting feather-like appearance. Popular varieties include - White City, Champion and Apricot.

- Thistle: The thistle bloom, also called the bush bloom, often features multi-colored blooms. The long, thin florets twist to rise up or fall backwards towards the stem. Thistle blooms have a unique, exotic look to them. Popular varieties include - Cindy, Cisco and Orange Spray.

Ornamental Uses

C. indicum.

Modern cultivated chrysanthemums are showier than their wild relatives. The flower heads occur in various forms, and can be daisy-like or decorative, like pompons or buttons. This genus contains many hybrids and thousands of cultivars developed for horticultural purposes. In addition to the traditional yellow, other colors are available, such as white, purple, and red. The most important hybrid is Chrysanthemum × morifolium (syn. C. × grandiflorum), derived primarily from C. indicum, but also involving other species.

Different colors of Chrysanthemum x morifolium.

Example of a Japanese bonsai chrysanthemum.

Over 140 cultivars of chrysanthemum have gained the Royal Horticultural Society's Award of Garden Merit. Chrysanthemums are divided into two basic groups, garden hardy and exhibition. Garden hardy mums are new perennials capable of wintering in most northern latitudes. Exhibition varieties are not usually as sturdy. Garden hardies are defined by their ability to produce an abundance of small blooms with little if any mechanical assistance, such as staking, and withstanding wind and rain. Exhibition varieties, though, require staking, overwintering in a relatively dry, cool environment, and sometimes the addition of night lights.

The exhibition varieties can be used to create many amazing plant forms, such as large disbudded blooms, spray forms, and many artistically trained forms, such as thousand-bloom, standard (trees), fans, hanging baskets, topiary, bonsai, and cascades.

Chrysanthemum blooms are divided into 13 different bloom forms by the US National Chrysanthemum Society, Inc., which is in keeping with the international classification system. The bloom forms are defined by the way in which the ray and disk florets are arranged. Chrysanthemum blooms are composed of many individual flowers (florets), each one capable of producing a seed. The disk florets are in the center of the bloom head, and the ray florets are on the perimeter. The ray florets are considered imperfect flowers, as they only possess the female reproductive organs, while the disk florets are considered perfect flowers, as they possess both male and female reproductive organs.

Irregular incurves are bred to produce a giant head called an *ogiku*. The disk florets are concealed in layers of curving ray florets that hang down to create a 'skirt'. Regular incurves are similar, but usually with smaller blooms and a dense, globular form. Intermediate incurve blooms may have broader florets and a less densely flowered head.

In the reflex form, the disk florets are concealed and the ray florets reflex outwards to create a mop-like appearance. The decorative form is similar to reflex blooms, but the ray florets usually do not radiate at more than a 90° angle to the stem.

The pompon form is fully double, of small size, and very globular in form. Single and semidouble blooms have exposed disk florets and one to seven rows of ray florets. In the anemone form, the disk florets are prominent, often raised and overshadowing the ray florets. The spoon-form disk florets are visible and the long, tubular ray florets are spatulate. In the spider form, the disk florets are concealed, and the ray florets are tube-like with hooked or barbed ends, hanging loosely around the stem. In the brush and thistle variety, the disk florets may be visible.

In Japan, a form of *bonsai* chrysanthemum was developed over the centuries. The cultivated flower has a lifespan of about 5 years and can be kept in miniature size. Another method is to use pieces of dead wood and the flower grows over the back along the wood to give the illusion from the front that the miniature tree blooms.

Culinary Uses

Yellow or white chrysanthemum flowers of the species C. morifolium are boiled to make a tea in some parts of Asia. The resulting beverage is known simply as chrysanthemum tea, pinyin: júhuā chá, in Chinese). In Korea, a rice wine flavored with chrysanthemum flowers is called gukhwaju.

Chrysanthemum leaves are steamed or boiled and used as greens, especially in Chinese cuisine. The flowers may be added to dishes such as mixian in broth, or thick snakemeat soup to enhance the aroma. Small chrysanthemums are used in Japan as a sashimi garnish.

Insecticidal Uses

Pyrethrum (Chrysanthemum [or Tanacetum] cinerariaefolium) is economically important as a natural source of insecticide. The flowers are pulverized, and the active components, called pyrethrins, which occur in the achenes, are extracted and sold in the form of an oleoresin. This is applied as a suspension in water or oil, or as a powder. Pyrethrins attack the nervous systems of all insects, and inhibit female mosquitoes from biting. In sublethal doses, they have an insect-repellent effect. They are harmful to fish, but are far less toxic to mammals and birds than many synthetic insecticides. They are not persistent, being biodegradable, and also decompose easily on exposure to light. Pyrethroids such as permethrin are synthetic insecticides based on natural pyrethrum.

Persian powder is an example of industrial product of chrysanthemum insecticide.

Environmental Uses

Chrysanthemum plants have been shown to reduce indoor air pollution by the NASA Clean Air Study.

Cultivation of Chrysanthemum

1. Climate Required for Chrysanthemum Cultivation: Basically , the chrysanthemum plant is a short day plant, in other words, it requires long days for vegetative growth and short days for flowering. Light and temperature are the main factors that influence the growth of the plants and flowering. For its vegetative growth it requires long day with bright sunlight and high temperature ranging from 20 °C to 28 °C. For bud formation and flowering it requires short day and low temperature ranging from 10 °C to 28 °C. This crop requires the relative humidity of 75% to 90% which suitable for proper plant growth.

2. Soil Requirement in Chrysanthemum Cultivation: The best suitable soil for chrysanthemum cultivation is well-drained sandy loam good textured soils. Having good amount of organic matter will result in excellent yield. Avoid the soils where too much of water stagnation is possible. The optimal soil pH range for its growth is 6.5 to 7.5.

3. Land Preparation in Chrysanthemum Cultivation: The land should be prepared by giving 2 to 3 ploughings followed by harrowing to prepare the beds for planting. Supplementing the farm yard manure of 20 to 25 tonnes/ha in the last plough is recommended.

4. Planting Season in Chrysanthemum Cultivation: Terminal cuttings of stock plants should be taken in the month of June and they should be transplanted after rooting in 15 cm pots at the end of July. These plants should be ready for pinching during end of Aug or beginning of Sept.

5. Propagation and Planting in Chrysanthemum Cultivation: Chrysanthemum plants are propagated vegetatively.

This various methods of propagation of chrysanthemum are:

Terminal Cuttings

These cuttings are taken from healthy stock plants from middle to end of June. 5-7 cm long cuttings are made by shearing basal leaves and cutting half of open leaves. For enhancing rooting, these cuttings can also be treated with Seradix-1 powder or 25 ppm NAA. Then these cuttings are planted in sand in pots or beds and are kept in partial shade.

Water is sprayed 4-5 times a day. Rooting takes place in 2-3 weeks and these cuttings are ready for transplanting. To avoid the rotting of cuttings, Captan (0.3%) or Brasicol (0.2%) should be applied in irrigation water once or twice.

Root Suckers

After flowering, the plants are planted in partial shade and ample of irrigation and fertilizer is applied for encouragement of root suckers. When these root suckers are 10-15 cm high, they are separated in February or March and planted in small pots and later on planted in bigger pots.

Seeds

This method of propagation is used by breeders to evolve new varieties. In majority of varieties seed set is less due to self-incompatibility and lack of pollination in winter by honey-bees. By artificial crossing seeds are obtained and sown for creating new variations.

Pinching and Disbudding

These are important operations which should be done carefully at appropriate time. The objective of pinching is to encourage the side branches and it should be repeated to encourage more number of branches depending upon the number of blooms to be retained.

On the contrary, disbudding is done to remove the side branches which arise from axillary buds so that number of flowers is limited and blooms of better size are obtained. If disbudding is not followed, many branches will be produced and bear the flowers which will affect the quality of blooms ultimately.

Standard Chrysanthemum

In this case single bloom on a branch is allowed to produce. The standard chrysanthemum Varieties have the genetic potential to produce single bigger sized bloom on a branch if disbudding and proper feeding is done. The pinching is not done if only one central bloom is desired on the main branch and only disbudding is done regularly in such cases.

Single pinching is done, if two flowers are desired whereas double pinching is done for four flowers. Similarly more number of pinching can be done to produce more standard flowers per plant. Regular disbudding is done to produce single flowers on single branch. First pinching is done in July and second in August and finally in September.

Spray Chrysanthemum

These types produce small to medium sized numerous flowers and do not have genetic potential to produce bigger size bloom irrespective of disbudding and best fertilization. These are planted for mass effect in beds or for commercial production.

In such varieties two pinching's are required to encourage lateral growth. First pinching is done at 4 weeks after planting and second after 7 weeks of planting. No or little disbudding is done. These spray chrysanthemum are also trained as cascades or in different shapes.

Staking

Staking is very necessary to provide the support whether plants are grown in pots or in field. There are only a few flowering varieties which neither require pinching nor staking and hence known as no stake no pinch varieties. For standard varieties, number of stakes will depend upon the number of main branches which have been allowed to produce bloom.

In spray type, 3-4 stakes are inserted on the band of the pot and are tied with string from bottom to top which gives good support to developing branches and finally flowers. When the plants are grown for cascade, the framework of steel or wooden strips is prepared first and plants are trained according to shape.

Growing of Chrysanthemum in Pots

Chrysanthemums are grown in pots to decorate verandahs, window gardens and to create colourful islands in the lawn. It provides a great pleasure in growing chrysanthemum in pots. Mostly earthen pots of different sizes are used. Now-a-days use of plastic pots is also on the increase.

The common pot mixture used is comprised of 3 parts garden soil: 1 part sand, 1 part leaf mould and 1 part of well rotten farmyard manure. Pots are filled by putting concave side of 3-5 pieces of crocks over the drain hole. For growing chrysanthemums in pots, plants are propagated by root suckers or terminal cuttings but latter method is preferred.

The terminal cuttings are got rooted by the method as described earlier. Rooted cuttings are transplanted in mid to end of July in small pots measuring 10 cm. The pots are watered twice depending upon the requirements. The plant grows sufficiently in a month and it needs repotting in bigger size pot measuring about 15 cm.

Final repotting is done in end of September when plants are transferred to 25-30 cm pot. Staking is done according to type of variety and method of training. Pinching and disbudding is also done according to type of variety and method of training as described earlier. Regular feeding with nitrogenous, phosphoric and potassic fertilizers in small doses is helpful in keeping plants healthy and to obtain good size blooms. Urea should not be applied as this is known to cause phytotoxicity.

At the time of appearance of buds, feeding with dilute solutions of cakes and superphosphate is highly beneficial in obtaining good size and shine of bloom. The use of growth retardants like phosphon or Alar is very effective in producing better size of blooms on dwarf plants. The soil drenching is done with Alar @ 33 g/100 L or phosphon at 10% liquid formulations normally at the rate of 100 ml of solution for 15 cm pot has been found effective.

Growing for Loose Flowers

The commercial cultivation of small flowered varieties of pompon or decorative group like Cameo, Baggi, Santi, Birbal Sahni etc. around the big cities has been found profitable. For raising commercial crop, a fresh crop is to be raised and grow every year. Soil is prepared in end of June by cultivating 2-3 times and adding of 10-15 cartloads of farmyard manure along with 100 kg single superphosphate and 133 kg of Muriate of Potash/acre before planting as basal dose.

The rooted cuttings prepared by the method described earlier are transplanted at the distance 30 × 20 cm in the evening hours of the day in middle of July. Irrigation is done immediately. In order to encourage the branching and ultimate flowering, pinching of apical portion of shoots should be done 4 and 7 weeks after transplanting. The application of 320 kg CAN/acre should be split into three doses and should be applied at the interval for 30 days.

Flowering occurs in November if planting has been done in July which continues up to six weeks. Flowering time can be extended by delay in planting up to end of August. Fully opened flowers are picked in late hours of day when dew has dried up. Soon after harvest, flowers are packed in baskets containing 3-4 kg of flowers. These varieties produce about 6000 kg of flowers/acre and are sold at a good price.

1. Manures and Fertilizers in Chrysanthemum Cultivation: Chrysanthemum crop responds very well to manuring and requires about 10 to 12 tonnes of well rotten farmyard manure (FYM) per acre. This farm yard manure can be supplemented at the time of land or soil preparation. As a basal dose, apply 50 kg of 'N' (Nitrogen), 160 kg P_2O_5 and 80 kg K_2O. For increasing the flower yield, spray GA_3 at 50 ppm at 30, 45 and 60 days after planting. Micronutrients like foliar spray of $ZnSO_4$ 0.25% + $MgSO_4$ 0.5% can be applied. As part of biofertilizers, soil application of 2 kg each of Azospirillum and Phosphobacteria per ha at the time of planting should be applied. It should be mixed with 100 kg of farmyard manure (FYM) and applied.

2. Irrigation in Chrysanthemum Cultivation: The frequency of irrigation depends on the stage of growth, soil and weather conditions. Proper drainage system should be maintained for chrysanthemum grown both in beds and in pots. Irrigation should be carried twice a week in the first month and subsequently at weekly intervals.

Chrysanthemum Flower Cultivation in Polyhouse

3. Weed Control in Chrysanthemum Cultivation: Weeds should be controlled or avoided for proper growth of the plants and good yields of flowers. If these flowers are grown in greenhouse, make sure to control the weeds as these weeds consume moisture and nourishment from plants. Two to three hand weeding should be carried out for proper growth of the plant. First weeding should be carried 4 weeks after planting. Good herbicide can also be applied to control weeds in greenhouse or open field or even in pots.

4. Pests and Diseases in Chrysanthemum Cultivation: Aphids, mites, thrips, leaf miners, leaf folder, root rot, leaf spot, wilt, rust, powdery mildew, chrysanthemum stunt and chrysanthemum mosaic disease are some common pests and diseases found in chrysanthemum cultivation. For control measures of these pests and diseases, contact floriculture department.

5. Flowering and Harvesting in Chrysanthemum Cultivation: Usually Chrysanthemum plants take 5 to 6 months from planting to flowering. It all depends on the cultivar (varieties), plant starts yielding flowers after 3 to 4 months of transplanting in the field. For cut-flower purpose, stem should be cut about 10 cm above the soil to avoid cutting into wooden tissue. The lower 1/3 of stem should be placed in water to extend the life of cut flowers. The best way to protect the flowers is to sleeve the flower bunch with a transparent plastic sleeve. The right stage for harvesting depends on the variety grown, marketing and purpose.

Ready to pick up chrysanthemum flowers

6. Post Harvesting in Chrysanthemum Cultivation: Loose flowers can be packed in bamboo baskets or gunny bags for marketing. The capacity of bamboo baskets ranges from 1 to 7 kg while gunny bags can accommodate 30 to 35 kg of loose flowers.

7. Yield in Chrysanthemum Cultivation: Generally, flowering seasons vary from region to region. The natural blooming seasons for most of the regions lasts from July to Feb. One can harvest the flowers around 15 times. The yield can be obtained from 10 to 11 tonnes of loose flowers per acre.

Lilium

Lilium longiflorum flower – 1. Stigma, 2. Style, 3. Stamens, 4. Filament, 5. Tepal

Lilium (members of which are true lilies) is a genus of herbaceous flowering plants growing from bulbs, all with large prominent flowers. Lilies are a group of flowering plants which are important in culture and literature in much of the world. Most species are native to the temperate northern hemisphere, though their range extends into the northern subtropics. Many other plants have "lily" in their common name but are not related to true lilies.

Lilies are tall perennials ranging in height from 2–6 ft (60–180 cm). They form naked or tunicless scaly underground bulbs which are their organs of perennation. In some North American species the base of the bulb develops into rhizomes, on which numerous small bulbs are found. Some species develop stolons. Most bulbs are buried deep in the ground, but a few species form bulbs near the soil surface. Many species form stem-roots. With these, the bulb grows naturally at some depth in the soil, and each year the new stem puts out adventitious roots above the bulb as it emerges from the soil. These roots are in addition to the basal roots that develop at the base of the bulb.

Lily, petal

The flowers are large, often fragrant, and come in a wide range of colors including whites, yellows, oranges, pinks, reds and purples. Markings include spots and brush strokes. The plants are late spring- or summer-flowering. Flowers are borne in racemes or umbels at the tip of the stem, with six tepals spreading or reflexed, to give flowers varying from funnel shape to a "Turk's cap". The tepals are free from each other, and bear a nectary at the base of each flower. The ovary is 'superior', borne above the point of attachment of the anthers. The fruit is a three-celled capsule.

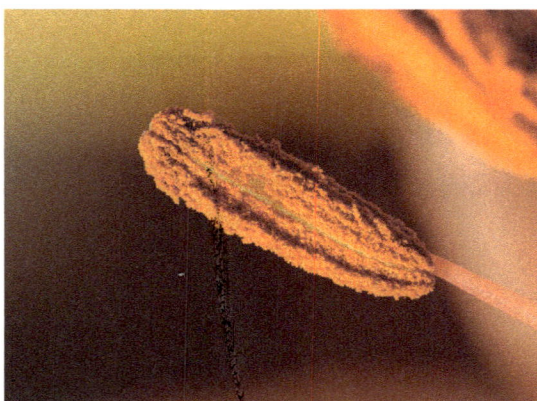
Stamen of lilium

Seeds ripen in late summer. They exhibit varying and sometimes complex germination patterns, many adapted to cool temperate climates.

Naturally most cool temperate species are deciduous and dormant in winter in their native environment. But a few species which distribute in hot summer and mild winter area (Lilium candidum, Lilium catesbaei, Lilium longiflorum) lose leaves and remain relatively short dormant in Summer or Autumn, sprout from Autumn to winter, forming dwarf stem bearing a basal rosette of leaves until, after they have received sufficient chilling, the stem begins to elongate in warming weather.

The basic chromosome number is twelve (n=12).

Lilium candidum seeds

Distribution and Habitat

The range of lilies in the Old World extends across much of Europe, across most of Asia to Japan, south to India, and east to Indochina and the Philippines. In the New World they extend from southern Canada through much of the United States. They are commonly adapted to either woodland habitats, often montane, or sometimes to grassland habitats. A few can survive in marshland and epiphytes are known in tropical southeast Asia. In general they prefer moderately acidic or lime-free soils.

Cultivation

Many species are widely grown in the garden in temperate and sub-tropical regions. They may also be grown as potted plants. Numerous ornamental hybrids have been developed. They can be used in herbaceous borders, woodland and shrub plantings, and as patio plants. Some lilies, especially Lilium longiflorum, form important cut flower crops. These may be forced for particular markets; for instance, Lilium longiflorum for the Easter trade, when it may be called the Easter lily.

Lilies are usually planted as bulbs in the dormant season. They are best planted in a south-facing (northern hemisphere), slightly sloping aspect, in sun or part shade, at a depth 2½ times the height of the bulb (except Lilium candidum which should be planted at the surface). Most prefer a porous, loamy soil, and good drainage is essential. Most species bloom in July or August (northern hemisphere). The flowering periods of certain lily species begin in late spring, while others bloom in late summer or early autumn. They have contractile roots which pull the plant down to the correct depth, therefore it is better to plant them too shallowly than too deep. A soil pH of around 6.5 is generally safe. The soil should be well-drained, and plants must be kept watered during the growing season. Some plants have strong wiry stems, but those with heavy flower heads may need staking.

Awards

The following lily species and cultivars currently hold the Royal Horticultural Society's Award of Garden Merit:

- African Queen Group (VI-/a) 2002 H6
- 'Casa Blanca' (VIIb/b-c) 1993 H6
- 'Fata Morgana' (Ia/b) 2002 H6
- 'Garden Party' (VIIb/b) 2002 H6
- Golden Splendor Group (VIb-c/a)
- Lilium henryi (IXc/d) 1993 H6
- Lilium mackliniae (IXc/a) 2012 H5
- Lilium martagon – Turk's cap lily (IXc/d)
- Lilium pardalinum – leopard lily (IXc/d)
- Pink Perfection Group (VIb/a)
- Lilium regale – regal lily, king's lily (IXb/a).

'Golden Splendor'

Classification of Garden forms

Numerous forms, mostly hybrids, are grown for the garden. They vary according to the species and interspecific hybrids that they derived from, and are classified in the following broad groups:

Asiatic Hybrids (Division I)

These are derived from hybrids between species in Lilium section Sinomartagon. They are derived from central and East Asian species and interspecific hybrids, including Lilium amabile, Lilium bulbiferum, Lilium callosum, Lilium cernuum, Lilium concolor, Lilium dauricum, Lilium davidii, Lilium × hollandicum, Lilium lancifolium (syn. Lilium tigrinum), Lilium lankongense, Lilium leichtlinii, Lilium × maculatum, Lilium pumilum, Lilium × scottiae, Lilium wardii and Lilium wilsonii.

These are plants with medium-sized, upright or outward facing flowers, mostly unscented. There are various cultivars such as Lilium 'Cappuccino', Lilium 'Dimension', Lilium 'Little Kiss' and Lilium 'Navona'. Dwarf (Patio, Border) varieties are much shorter, c.36–61 cm in height and were designed for containers. They often bear the cultivar name 'Tiny', such as the 'Lily Looks' series, e.g. 'Tiny Padhye', 'Tiny Dessert'.

Martagon Hybrids (Division II)

These are based on Lilium dalhansonii, Lilium hansonii, Lilium martagon, Lilium medeoloides, and Lilium tsingtauense.

The flowers are nodding, Turk's cap style (with the petals strongly recurved).

Candidum (Euro-Caucasian) Hybrids (Division III)

This includes mostly European species: Lilium candidum, Lilium chalcedonicum, Lilium kesselringianum, Lilium monadelphum, Lilium pomponium, Lilium pyrenaicum and Lilium × testaceum.

American Hybrids (Division IV)

These are mostly taller growing forms, originally derived from Lilium bolanderi, Lilium × burbankii, Lilium canadense, Lilium columbianum, Lilium grayi, Lilium humboldtii, Lilium kelleyanum, Lilium kelloggii, Lilium maritimum, Lilium michauxii, Lilium michiganense, Lilium occidentale, Lilium × pardaboldtii, Lilium pardalinum, Lilium parryi, Lilium parvum, Lilium philadelphicum, Lilium pitkinense, Lilium superbum, Lilium ollmeri, Lilium washingtonianum, and Lilium wigginsii.

Many are clump-forming perennials with rhizomatous rootstocks.

Longiflorum Hybrids (Division V)

These are cultivated forms of this species and its subspecies.

They are most important as plants for cut flowers, and are less often grown in the garden than other hybrids.

Trumpet Lilies (Division VI), including Aurelian Hybrids (with L. Henryi)

This group includes hybrids of many Asiatic species and their interspecific hybrids, including Lilium × aurelianense, Lilium brownii, Lilium × centigale, Lilium henryi, Lilium × imperiale, Lilium × kewense, Lilium leucanthum, Lilium regale, Lilium rosthornii, Lilium sargentiae, Lilium sulphureum and Lilium × sulphurgale.

The flowers are trumpet shaped, facing outward or somewhat downward, and tend to be strongly fragrant, often especially night-fragrant.

Oriental Hybrids (Division VII)

These are based on hybrids within Lilium section Archelirion, specifically Lilium auratum and Lilium speciosum, together with crossbreeds from several species native to Japan, including Lilium nobilissimum, Lilium rubellum, Lilium alexandrae, and Lilium japonicum.

They are fragrant, and the flowers tend to be outward facing. Plants tend to be tall, and the flowers may be quite large. The whole group are sometimes referred to as "stargazers" because many of them appear to look upwards.

Other Hybrids (Division VIII)

Includes all other garden hybrids.

Species (Division IX)

All natural species and naturally occurring forms are included in this group.

The flowers can be classified by flower aspect and form:

1. Flower Aspect:

- Up-facing

- Out-facing

- Down-facing

2. Flower Form:

- Trumpet-shaped

- Bowl-shaped

- Flat (or with tepal tips recurved)

- Tepals strongly recurved (with the Turk's cap form as the ultimate state).

Many newer commercial varieties are developed by using new technologies such as ovary culture and embryo rescue.

Pests and Diseases

Scarlet lily beetles, Oxfordshire, UK

Aphids may infest plants. Leatherjackets feed on the roots. Larvae of the Scarlet lily beetle can cause serious damage to the stems and leaves. The scarlet beetle lays its eggs and completes its life cycle only on true lilies (Lilium) and fritillaries (Fritillaria). Oriental, rubrum, tiger and trumpet lilies as well as Oriental trumpets (orienpets) and Turk's cap lilies and native North American Lilium species are all vulnerable, but the beetle prefers some types over others. The beetle could also be having an effect on native Canadian species and some rare and endangered species found in northeastern North America. Daylilies (Hemerocallis, not true lilies) are excluded from this category. Plants can suffer from damage caused by mice, deer and squirrels. Slugs, snails and millipedes attack seedlings, leaves and flowers. Brown spots on damp leaves may signal botrytis (also known as lily disease). Various fungal and viral diseases can cause mottling of leaves and stunting of growth.

Propagation and Growth

Lily growing in Eastern Siberia

Lilies can be propagated in several ways;

- By division of the bulbs.

- By growing-on bulbils which are adventitious bulbs formed on the stem.

- By scaling, for which whole scales are detached from the bulb and planted to form a new bulb.

- By seed; there are many seed germination patterns, which can be complex.

- By micropropagation techniques (which include tissue culture); commercial quantities of lilies are often propagated in vitro and then planted out to grow into plants large enough to sell.

According to a study done by Anna Pobudkiewicz and Jadwiga the use of flurprimidol foliar spray helps aid in the limitation of stem elongation in oriental lilies.

Toxicity

Some Lilium species are toxic to cats. This is known to be so especially for Lilium longiflorum though other Lilium and the unrelated Hemerocallis can also cause the same symptoms. The true mechanism of toxicity is undetermined, but it involves damage to the renal tubular epithelium (composing the substance of the kidney and secreting, collecting, and conducting urine), which can cause acute renal failure. Veterinary help should be sought, as a matter of urgency, for any cat that is suspected of eating any part of a lily – including licking pollen that may have brushed onto its coat.

Culinary and Herb uses

China

Lilium bulbs are starchy and edible as root vegetables, although bulbs of some species may be very bitter. The non-bitter bulbs of Lilium lancifolium, Lilium pumilum, and especially Lilium brownii and Lilium davidii var. unicolor are grown on a large scale in China as a luxury or health food, and are most often sold in dry form for herb, the fresh form often appears with other vegetables. The dried bulbs are commonly used in the south to flavor soup. Lily flowers are also said to be efficacious in pulmonary affections, and to have tonic properties. Lily flowers and bulbs are eaten especially in the summer, for their perceived ability to reduce internal heat. They may be reconstituted and stir-fried, grated and used to thicken soup, or processed to extract starch. Their texture and taste draw comparisons with the potato, although the individual bulb scales are much smaller. There are also species which are meant to be suitable for culinary and/or herb uses.

There are five traditional lily species whose bulbs are certified and classified as "vegetable and non-staple foodstuffs" on the National geographical indication product list of China.

- Culinary Use: Lilium brownii, Lilium brownii var. viridulum, Lilium concolor, Lilium dauricum, Lilium davidii, Lilium distichum, Lilium lancifolium, Lilium martagon var. pilosiusculum, Lilium pumilum, , Lilium speciosum var. gloriosoides.

- Herb Use: Lilium brownii, Lilium brownii var. viridulum, Lilium concolor, Lilium dauricum, Lilium lancifolium, Lilium pumilum, Lilium rosthornii, Lilium speciosum var. gloriosoides, Lilium sulphureum.

And there are researches about the selection of new varieties of edible lilies from the horticultural cultivars, such as 'Batistero' and 'California' among 15 lilies in Beijing, and 'Prato' and 'Small foreigners' among 13 lilies in Ningbo.

The "lily" flower buds known as jīnzhēn "golden needles" in Chinese cuisine are actually from Hemerocallis citrina.

Japan

- Culinary Use: Yuri-ne (lily-root) is also common in Japanese cuisine, especially as an ingredient of chawan-mushi (savoury egg custard). The major lilium species cultivated as vegetable are Lilium leichtlinii var. maximowiczii, Lilium lancifolium, and Lilium auratum.

- Herb Use: Lilium lancifolium, Lilium brownii var. viridulum, Lilium brownii var. colchesteri, Lilium pumilum

North America

The flower buds and roots of Lilium canadense are traditionally gathered and eaten by North American indigenous peoples. Coast Salish, Nuu-chah-nulth and most western Washington peoples steam, boil or pit-cook the bulbs of Lilium columbianum. Bitter or peppery-tasting, they were mostly used as a flavoring, often in soup with meat or fish.

Taiwan

- Culinary Use: The parts of lilium species which are officially listed as food material are the flower and bulbs of Lilium lancifolium Thunb., Lilium brownii var. viridulum Baker, Lilium pumilum DC., Lilium candidum Loureiro. Most edible lily bulbs which can be purchased in a market are mostly imported from mainland China (only in the scale form, and most marked as Lilium davidii var. unicolor) and Japan (whole bulbs, should mostly be Lilium leichtlinii var. maximowiczii). There are already commercially available organic growing and normal growing edible lily bulbs. The varieties are selected by the Taiwanese Department of Agriculture from the Asiatic lily cultivars that are imported from the Netherlands; the seedling bulbs must be imported from the Netherlands every year.

- Herb Use: Lilium lancifolium Thunb., Lilium brownii var. viridulum Baker, Lilium pumilum DC.

South Korea

- Herb Use: The lilium species which are officially listed as herbs are Lilium lancifolium Thunberg; Lilium brownii var. viridulun Baker.

Marigold

Marigolds are hardy, annual plants and are great plants for cheering up any garden. Broadly, there are two genuses which are referred to by the common name, Marigolds viz. Tagetes and Celandula. Tagetes includes African Marigolds and French Marigolds. Celandula includes Pot Marigolds.

Marigolds come in different colors, yellow and orange being the most common. Most of the marigolds have strong, pungent odor and have great value in cosmetic treatment. There are many varieties of Marigolds available today. Some of the major Marigold varieties are listed below:

- African or American Marigolds (Tagetes erecta): These marigolds are tall, erect-growing plants up to three feet in height. The flowers are globe-shaped and large. Flowers may measure up to 5 inches across. African Marigolds are very good bedding plants. These flowers are yellow to orange and do not include red colored Marigolds. The Africans take longer to reach flowering stage than the French type.

- French Marigolds (Tagetes patula): Marigold cultivars in this group grow 5 inches to 18 inches high. Flower colors are red, orange and yellow. Red and orange bicolor patterns are also found. Flowers are smaller (2 inches across). French Marigolds are ideal for edging flowerbeds and in mass plantings. They also do well in containers and window boxes.

- Signet Marigolds (T. signata 'pumila'): The signet Marigolds produce compact plants with finely divided, lacy foliage and clusters of small, single flowers. They have yellow to orange colored, edible flowers. The flowers of signet marigolds have a spicy tarragon flavor. The foliage has a pleasant lemon fragrance. Signet Marigolds are excellent plants for edging beds and in window boxes.

- Mule Marigolds: These marigolds are the sterile hybrids of tall African and dwarf French marigolds, hence known as mule Marigolds. Most triploid cultivars grow from 12 to 18 inches high. Though they have the combined qualities of their parents, their rate of germination is low.

Cultivation of Marigold

Soil and Climate Requirements of Marigold

Soil

Marigold can be successfully cultivated on a wide variety of soils. However, a soil that is deep fertile, friable having good water holding capacity well drained and near to neutral in reaction viz. pH 7.0-7.5 is most desirable.

Climate

Marigold requires mild climate for luxuriant growth and profuse flowering. It ceases to grow at high temperature thereby flower quantity and quality is adversely attested. During severe winter including frost plants and flowers are killed and blackened.

However, plants if allowed to over winter, sprout during spring season and produce some flowers. Sowing and planting is carried out during rainy season, winter and summer season. Hence, flowers of marigold can be had almost throughout the year.

Under north Indian conditions in the plains the best flowering has been observed during winter months i.e., from October to April. During summer which is characterized by high temperature and long days, plants remain mostly vegetative with sparse flowering. Under such conditions, flower size is reduced considerably but length of flower stalk is increased. On the contrary in the hills flowering commences from May onwards which continues up to October.

Preparation of Soil

Land should be well prepared by ploughing it 2-3 times and 50 tons of well rotten farmyard manure should be well mixed/ha. Beds of convenient size are made to facilitate irrigation and other cultural operations.

Planting

There are two common methods of propagation of marigold i.e., (i) by seeds, (ii) by cuttings. Crop raised from seeds is tall, vigorous and heavy bloomer; thus it is preferred over cuttings.

Seed Rate and Nursery Rising

Seeds of wide range of varieties of common species i.e., T. erecta, T. patula, T. tenuifolia are easily available and germinate quickly. Therefore, propagation through seed is advised. For better seed germination, optimum temperature range is between 18 to 30 °C. For raising seedlings for one hectare, about 1.5 kg seed is required.

Seeds of marigold can be sown in pots, seed boxes or on flat or raised nursery beds. Nursery beds of 3 × 1 m size are thoroughly prepared and mixed with 10 kg of well rotten farmyard manure per sq m. About 8 to 10 such beds are needed to raise seedlings for one hectare. Before sowing the seeds, D.D.T. or B.H.C. should be dusted on outer side of nursery beds to avoid removal of seeds by ants.

Seeds can be sown preferably in lines or by broadcast method. In case of broadcasting care should be taken for proper distribution of seeds so as to have healthy seedlings. For this, thinning is an important operation in both the methods. Seeds need to be covered with light soil on sand or strained leaf mould and watering should be done with rose can. For entire period, nursery bed should be kept moist but not wet and thus watered accordingly.

Sowing Time

Marigold Crop Can be raised three times a year i.e., rainy, winter and summer season.

Sowing and Planting times for Each Season are as under:

Season	Sowing time	Transplanting time
Rainy Winter Summer	End of June to 1st week of July Mid of September First week of January (Under glass house or plastic)	First fortnight of August Mid of October First week of February

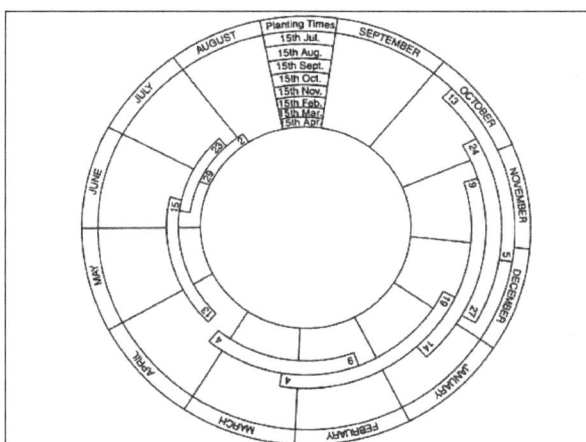

Figure: Showing the period of Availability of Flowers as affected by Planting Times.

A scheme has been developed to produce marigold flowers from October to May- June. By following the growers around the cities can produce flowers for longer duration and good profit can be earned. Performance of a variety varies from season to season which is reflected in the data given in table for Giant Double African Orange. Size of flowers is likely to vary within a season whether flowers plucked or not.

Cuttings

This method is commonly followed for maintaining the purity of varieties. Normally, the presence of adventitious roots along the stem helps in the establishment of cuttings. About 10 cm long cutting are made and treated with seradix No. 1. The cuttings are planted in the sand to strike roots easily and plants thus raised are used for bedding and pot planting.

Table: Effect of planting seasons on growth and flowering marigold (T. erects) cv. Giant Double African Orange.

Season	Height (em)	Days required for first flowering	No of flowers plant(not plucked) (not	No of flowers plant(plucked regularly)	Yield of flower (kg/ plant)(plucked regularly)	Flower size (cm)
Rainy	98.5	58	35	54	0.492	8.20
Winter	59.9	53	21	32	0.382	9.05
Summer	35.5	58	6	12	0.042	4.55

Transplanting of Seedlings

Marigold seedlings are easily established after transplanting in the field without much mortality. Fast growing root system presents in this species enable the seedlings to establish better. At the time of transplanting, seedlings normally of about one month old should have attained 3-4 true leaves and must be stocky.

Closely sown seeds give rise to thin, long and weak plants which are leggy and do not make a good plant. Very old seedlings are also not desirable because they have lost their juvenile phase in the nursery itself. Seedlings should be transplanted in well prepared land in the evening especially during rainy and summer seasons to avoid exposure to harsh weather, transplanting shock and to allow better establishment in cool hours of night.

Soil should be pressed well around root zone so that there are no air spaces left. After transplanting, a light irrigation is essential. Heavy irrigation if done immediately after transplanting seedlings bent down and leaves stick with the soil which delays proper development of plant. On light sandy soil irrigation a day before transplanting is considered beneficial for the better establishment of seedlings.

Spacing

Proper spacing between plants is required for growth, development and flower production. Two marigold species which are commonly grown for commercial production of flowers require different spacing's, for example Targets erecta requires wider spacing's than T. patula. Experiments conducted at RA.U. reveal that high flower production in Giant Double African Orange (T. erecta) was obtained by planting at a distance of 40 x 30 cm whereas in case of T. patula cv. Red Brocade higher number and weight of flowers per plant was recorded in spacing's 20 × 20 cm; however, per unit area, highest yield obtained was from 20 × 10 cm spacing.

Manures and Fertilizers for Marigold

Although marigold is one of the commercial flowers but much research work has not been done to determine its requirements for macro and micro nutrients for growth and flower production. The rate of nutrient utilization from soil has been reported to be 57.5 per cent of available N, 20.1 per cent R 94.4 per cent K, 8.6 per cent Ca, 23.0 per cent Mg and 6.4 per cent Na.

Field trials were conducted at RA.U. for two years to find out optimum dose of N and P of marigold (Tagetes erecta) cv. Giant Double African Orange for growth and' flower production. The indication was that with the increase of N from 0 to 40 g/sq m, the number of flowers per plant and flower yield increased with the increase in dose whereas plant height responded only up to 30 g/sq m.

The response to P was inconsistent on growth and flower production. Therefore, to get highest flower yield, 100 kg N (4 Q CAN) 100 kg P_2O_5 (6.25 Q single superphosphate) and 100 kg K_2O per ha should be mixed at the time of preparation of land. Remaining 100 kg N (4 Q CAN) per ha should be applied one month after seedlings are transplanted.

Weeding and Hoeing

Weeds are a problem in marigold especially in rainy season crop. After transplanting of marigold seedlings in the field, weeds grow faster than marigold in initial stages and cover large area in a few days. If these weeds are not removed in time, a great loss would occur in terms of growth and productivity of marigold. In India, weeding is done manually. It has been observed that 3-4 times weeding's are required during the entire growth period.

Irrigation

Marigold is a herbaceous plant and puts up rapid vegetative growth during initial stages. It takes about 55-60 days to complete vegetative growth and to enter into reproductive phase. At this stage terminal flower buds appear which break the apical dominance and encourage secondary branches.

At all stages of vegetative growth and during flowering period sufficient amount of moisture in soil is essential. Moisture stress at any stage of growth and development may hamper the normal growth and productivity of flowers. The frequency and quantity of water mainly depends upon soil and climatic conditions.

In lighter soil more frequent irrigation is required than that in heavy soil. Season of planting determines the frequency of irrigation. If rainfall is normal and well distributed, irrigation is not frequently required, but if the rain is scanty, irrigation is needed frequently.

It has been experienced under sandy loam conditions of Punjab that from September to March, weekly irrigation is enough. Though "plants tolerate dry weather up to 10 days without irrigation but growth and flower production is affected adversely. From April to June, frequent irrigation at the interval of 4-5 days is required.

Pinching of Marigold Plants

In tall varieties of Tagetes erecta emergence of side branches and their flowering is influenced by the presence of apical dominance. It has been observed that the plants of marigold grow straight upwards to their final height and develop into terminal flower bud. By production of terminal flower bud, side buds become free from correlative inhibition of apical dominance and these buds develop into branches to produce flowers.

However, if the terminal portion of shoot is removed early, emergence of side branches starts earlier and more number of flowers of good quality and uniform size are produced. Results of studies on pinching revealed that pinching the plants 40 days after transplanting enabled the plants to yield more flowers. However, the plants remained dwarf and flowering was delayed in comparison to late pinching i.e., 50 or 60 days after transplanting.

Harvesting, Packing and Transportation of Marigold

Harvesting of flowers

Marigold flowers should be plucked when they attain the full size depending upon the variety. Plucking of flowers should be done in cool hours of the day i.e., either in the morning or evening. Field should be irrigated before so that flowers keep well of longer period after plucking.

Plucking is done by hand because flower stalk is a hollow structure which breaks easily when twisted between thumb and finger. Productivity of plants is increased considerably by regular plucking of flowers. It has been observed that in an acre flowers can be plucked by 6-8 labourers in a day. Flowers plucked should be covered with moist gunny bags if kept over-night before taking to market.

Packing

Marigold flowers are primarily used for making garlands and hence plucked flowers are collected in gunny bags or bamboo baskets for carrying to the market. For the local market marigold flowers are taken into gunny bags whereas for distant market bamboo baskets are used.

Transportation

Different means of transportation including rickshaws, buses and trains are used to carry the flowers to market depending upon the distance.

Yield of flowers

Flower yield depends upon season of planting and cultural practices adopted. On an average a fresh flower yield of 200-225 q per ha during rainy season, 150 to 175 q per ha in winter and 100-120 q/ha in summer can be obtained.

Seed Production

Marigold is a cross pollinated crop, hence, proper isolation distance of 1-1.5 km should be given amongst varieties. However, natural cross pollination amongst species is absent. The seed crop is ready in about six months.

Tuberose

Tuberose (Polianthes tuberosa L.) is one of the most important tropical ornamental bulbous flowering plants cultivated for production of long lasting flower spikes. belongs to the family Amaryllidaceae and is native of Mexico. Tuberose is an important commercial cut as well as loose flower crop due to pleasant fragrance, longer vase-life of spikes, higher returns and wide adaptability to varied climate and soil. They are valued much by the aesthetic world for their beauty and fragrance. The flowers are attractive and elegant in appearance with sweet fragrance. It has long been cherished for the aromatic oils extracted from its fragrant white flowers. Tuberose blooms throughout the year and its clustered spikes are rich in fragrance; florets are star shaped, waxy and loosely arranged on spike that can reach up to 30 to 45 cm in length. The flower is very popular for its strong fragrance and its essential oil is important component of high- grade perfumes. 'Single' varieties are more fragrant than 'Double' type and contain 0.08 to 0.14 percent concrete which is used in high grade perfumes. There is high demand for tuberose concrete and absolute in international markets which fetch a very good price. Flowers of the Single type (singlerow of perianth) are commonly used for extraction of essential oil, loose flowers, making garland etc., while that of Double varieties (more than two rows of perianth) are used as cut flowers, garden display and interior decoration. Fragrance of flowers is very sweet, floral and honey-like and can help give emotional strength. The flower spike of tuberose remains fresh for long time and finds a distinct place in the flower markets.

Importance and Uses

Tuberose can successfully be grown in pots, borders, beds and commercially cultivated for its various uses. The flowers of tuberose are also used for making artistic garlands, floral ornaments,bouquets, buttonholes, gajras and extraction of essential oil. It is also a popular cut flower, not only for use in arrangements, but also for the individual florets that can provide fragrance to bouquets and boutonnieres. The long flower spikes are excellent as cut flowers for table decoration. The flowers emit a delightful fragrance. Tuberose represents sensuality and is used in aromatherapy for its ability to open the heart and calm the nerves, restoring joy, peace and harmony. Tuberose flowers have long been used in perfumery as a source of essential oils and aroma compounds. Tuberose oil is used in high value perfumes and cosmetic products. Furthermore, fragrant flowers are added along with stimulants or sedatives to the favourite beverage prepared from chocolate and served either cold or hot as desired. Tuberose bulbs contain an alkaloid -lycorine, which causes vomiting. The bulbs are rubbed with turmeric and butter and applied as a paste over red pimples of infants. Dried tuberose bulbs in powdered form are used as a remedy for gonorrhoea. In Java, the flowers are eaten along with the juices of the vegetables.

Species and varieties

There are about fifteen species under the genus Polianthes, of which twelve species have been reported from Mexico and Central America.Of these, nine species have white flowers, one is white tinged with red and two are red. Except Polianthes tuberosa L., all the others are found growing wild.

Varieties

There are four types of tuberoses named on the basis of the number of rows of petals they bear. They are,

- Single

- Semi-double

- Double and

- Variegated.

Cultivation of Tuberose

Climate

Tuberose is a half hardy bulbous crop and happily grows in mild climate. It can tolerate wide range of climatic conditions and areas having a range of 20-35 °C temperature are considered good for its successful growing. For its luxuriant growth and high productivity the mild temperature of about 30 °C with high humidity is most ideal.

The temperature above than 40 °C reduces the productively and quality of spikes. Much lower temperature i.e. 1-2 °C with ground frosts kill the vegetative parts of the crop completely. However, sprouting and flowering is normal in ratoon crop with the onset of spring.

Soil

Tuberose is a bulbous plant; hence, loamy and sandy loam soils having pH of 6.5 to 7.5 with good aeration and drainage are highly pertinent for its successful cultivation. It has been observed that it can tolerate higher degree of pH i.e. up to 8.2 to 8.3. Its cultivation can be extended in marginal and unproductive soils affected by salinity and alkalinity. The soil should be rich in organic matter and good water holding capacity for proper growth of tuberose.

Preparation of Field

The field should be well prepared by giving 2-3 ploughings and incorporating 40-50 tonnes/ha of well rotten farm yard manure before planting. The application of 100 kg N and 400 kg P_2O_5 and 200 kg K_2O/ha is found to be optimum which should be applied before planting the crop along with 8 kg of thimet or 2.5 kg carbofuron/ha should be mixed to protect the crop of tuberose against nematode.

Bulb Selection and Planting

The size of bulb, its diameter and weight reflects the storage of food material which will be utilized for the growth, flower production and subsequent production of new bulbs. In tuberose, the production of daughter bulbs and flower production go simultaneously and hence, sufficiently good sized bulbs should be used for planting the crop. On an average the bulb size of 2-3 cm is good for raising healthy crop. But relatively larger bulb i.e. 3.1-3.5 cm weighing about 40-50 g exhibit better result and produced good quality spikes.

After field preparation, the field is divided into plots leaving the path and water channels. Bulbs are planted in the field at the distance of 20 × 20 cm and at 4-5 cm deep. Little more wider distance of 30 × 30 cm is followed in Maharashtra. Close planting at the distance of 15 x 15 cm has been found satisfactory at many locations including Assam. After planting field is irrigated lightly.

Manuring

Tuberose is heavy feeder and very exhaustive crop and thus responds well to the applied organic and inorganic nutrients. Fertilizer application of 200, 400 and 200 kg N, P_2O_5 and K_2O per ha, respectively is recommended for quality flower production and bulb production.

Half of N and full dose P and K to be applied before planting and remaining half dose of N to applied one month after planting. This can be increased or decreased depending upon fertility of soil. From Assam, it has been reported that the application of 80-40-60 g/m^2 of N, P_2O_5 and K_2O in plant crop and ratoon crop was found highly beneficial in terms of flower production and bulb production of single variety of tuberose.

Irrigation

Tuberose requires sufficient moisture for its growth and quality flower production. It is advised that a light irrigation should be given before planting which helps in early sprouting and growth of plant. Too much moisture at the time of sprouting of bulb result into rotting of bulbs. Similar conditions during flowering affect the quality of spikes adversely.

The frequency of irrigation depends upon soil types, weather conditions and stage of growth. In general weekly irrigation is required during April-June. Irrigation interval during July-September depends upon rains and their intensities whereas during October to March irrigation at the interval of 10 days keeping the rains in view is deficient to meet its requirement.

Weeding

It is advised to keep the tuberose field free from weed otherwise they compete for nutrient and moisture with the main crop. Since the tuberose is in the field for full 1-3 years and different kind of weeds appear season wise.

Hand weeding is laborious, time consuming and expensive. Generally after each irrigation, weeding is done. Hoeing is also done to loose soil and destroys the weeds. Pre-planting application of atrazine @ 3.0 kg/ha-has been found effective in controlling weeds in tuberose.

Harvesting, Packing and Marketing of Tuberose

Depending upon the purpose and the use of tuberose flowers, harvesting is done accordingly. For loose flowers, individual flower is plucked regularly which are used for various purposes whereas for cut flower the spike is cut from the base so that longer spike is available. These operations are done in cool hours either early in the morning or in the evening keeping the harvest of spike taken for the travel of these spikes.

Similarly the stage of harvest of spike is decided upon distance of travel. For local market and immediate purpose spikes are harvested when two-three rows of basal flowers are open. For distant market, spikes are harvested when basal flower just going to open. Loose flowers are generally packed in bamboo baskets, holding about 10-15 kg flowers and are sent to the market. These are sold by weight.

The quality of cut spikes are judge according to the length of spikes, length of flowering zone, number of florets, quality of florets. These cut spike are bunched in round bundles of 100 spikes. Basal portion of the spikes wrapped with newspaper. To avoid any injury to flowers and buds, the whole bundles is wrapped in soft, white tissue paper or perforate polythene of 300 gauges.

A holding solution of 2% sucrose + 300 ppm aluminum sulphate was found best for the post-harvest life and quality of cut spikes of tuberose cv Shringar. Another useful combination is 1% sucrose + 300 ppm citric acid + 50 ppm calcium nitrate at 4 °C for 48-72 hours to improve the vase life of spikes. These spikes can be stored at 4°-7° for 5-7 days.

Lifting and Curing of Bulbs

At the time of complete maturity growth ceases completely, leaves become yellow and dry out. At this stage irrigation is withheld and after two weeks, clumps of bulbs are dug out from the ground. The bulbs are separated out and grouped according to their size. Cleaned and graded bulbs are stored on shelves for curing.

The position of these bulbs is changed after two weeks to prevent fungal attack and rotting. Then bulbs are stored in cool and shady place at about 18 °C for four weeks before planting which result in production of good quality of bulb and flower spikes of subsequent crop. Longer storage at 30 °C or above before planting advances flowering but quality of spikes and bulbs deteriorates.

Flower and Concrete Yield

Flower production depend upon climatic conditions, soil type, its fertility level and agronomic practices followed, etc. Under optimum conditions single tuberose yields about 5,00,000 flower spikes/ha or 10.5 tonnes/ha of loose flower. In a cycle of three years, first two years flower yield is high and in third year it is comparatively lower. In one year about 9.2 to 10 kg concrete/ha can be obtained and in turn 1.75 kg of tuberose absolute can be obtained.

Method of Extraction of Tuberose

The simple method of steam distillation for extracting the volatile oil of tuberose cannot be adopted as tuberose do not yield the essential oil on steam distillation. Hence, the solvent extraction method is employed. Freshly plucked fully opened flowers in cool hours of the day (before 8.00 a.m.) are used.

The purified hexane is a colourless liquid with a boiling point of 60-80 °C. The flowers are immersed in purified hexane for 30 minutes. The complete extraction of the perfume from the flower can be ensured by slowly rotating the extractor having flowers and solvent for about 20 minutes. After the complete extraction of the perfume from the flowers, the solvent is filtered and concentrated at a constant temperature at 70 °C.

At this temperature solvent evaporates and condensed for reuse, leaving the perfume and other plant waxes. To purify further the remaining vapours of solvent containing perfume, wax and pigments is distilled in a vacuum distillation unit where the complete removal of solvent takes place, leaving floral concrete.

In any unit, four Components are there:

(1) Extractor,

(2) Evaporator,

(3) Condenser and

(4) Vacuum distillation.

Important Insects, Pests and Diseases of Tuberose:

Aphids

These are very small size insects feed on flower buds and on growing tips. These can be controlled by spraying 0.1 per cent spray Malathion or Rogor.

Thrips

These feed on leaves, flowers stalks and flowers and suck the sap and ultimately damage the crop. These can be controlled by spraying 400 ml of Rogor. 30 EC (dimethoate) in 80-100 L of water per area when the attack starts and repeat after 15 days if necessary.

Red Spider Mites

Yellow stripes and streaks appear on the foliage due to attack of mites. Spraying with endrin 9 kg of 10% wp/100 gal of water has been effective in controlling mites.

Root Knot Nematode

It is caused by Meloidygyne incognita, and has been observed that root knot nematodes reduce crop yield considerably and make the plants highly susceptible to the attack of Fusarium oxysporium sp. dianthi. Various methods like crop rotation, treatment of bulb before planting (dip. bulbs in 0.5% nuvacron 40 E.C. for half an hour) or treatment of soil with Basamid @ 40 g/m² or addition of neem leaves in the soil is recommended.

Recently it has been reported from IIHR, Bangalore that neem based formulation of Pochonia chlamydosporia, (verticillium chalamydosporium) (40 g @ 10⁶ CFU/g) and Trichoderma harzianum (40 g @ 10⁶ CFU/g) per 2 × 2 m sized plots before planting and subsequently once in 4 months decreased the incidence of root knot nematode and wilt disease complex significantly. These treatments increased the number of flotes per spike and spikes/plot, significantly.

Application of 8 kg thimet or neem cake @ 1 tonne/ha controls nematode infestation. Besides, the use of nematode free planting material and hot water treatment of bulbs at 57-58 °C for 30 minutes is recommended. Further, the population of nematode in the field can be kept under check when intercropping with marigold is followed.

Diseases of Flowers and their Management

Diseases of Rose - Powdery Mildew Sphaerotheca Pannosa var. Rosae

Symptoms: The disease appears as slightly raised blister-like areas on the young leaves. Soon leaves are covered with a greyish white, powdery fungal growth, become curled and distorted. On older leaves, large white patches of fungal growth appear. Buds may also be attacked and covered with white mildew before they open. Diseased buds fail to open. The infection spreads to theflower parts and they become discoloured, dwarfed and dried.

Favourable conditions: The fungus overwinters as mycelium in dormant buds and shoots spread is through wind-borne conidia. The disease is favoured by dry weather with maximum day temperature of 20 to 25 °C with cool nights.

Management

- The diseased and fallen leaves should be collected and burnt.

- Four sprayings at 10 days interval with wettable sulphur 0.3 per cent or dinocap 0.07 per cent or carbendazim 0.1 per cent or Azoxystrobin @ 1ml/ litre controls the disease effectively.

- Sulphur dust can be used at 25 kg/ha. Wettable sulphur or sulphur dust should not be used when the temperature is above 30 °C as it may cause scorching.

- Resistant Varieties like Aawliver, Abisharika, Adolf Morstman, African Star, Ambika, Angeles, Anvil Sparks, American Pride, Apollo, Arizona, Ashwini, Baby Masquerade, Banjaran-9, Barbara, Bewitched, Blue Moon, Bon Soir, Bon Accord, Bonnie Scotland, Boque Dayal, Belle Vue, Bovinchor, Bulls Red, Canasta, Careless Love, Carcusar, Celebration, Crimson Glory, Dame De Cour, Deep Secret, Dutch Gold, Dwarf Queen, Dearest Durina, Eiffel Tower, Priti, Paradise, Queen Elizabeth, Royal Ascot, Red Master Piece, Red Dene, Rachel Grawshey, Ranjana-1 0, Sonia, Spartan, Super Star, Summer Days, Starina can be grown. Excess fertilization especially with nitrogenous fertilizers and crowding of plants should be avoided.

Die-back- Diplodia Rosarum

Symptoms: The pruned surface of the twig dries from tip downwards. Twigs become brown to dark brown or black. The disease passes from the branch twig to the main stem and from where it spreads to the roots. Finally it kills the whole plant. Stem and roots show browning of the internal tissues.

Favourable conditions: The fungus persists in dead twigs and the stalks of the withered blooms. Older plants and neglected and weak bushes are more frequently attacked. Disease spread is faster at 30 to 32 °C.

Management

- Pruning should be done so that the lesions on the shoots are eliminated. Partially diseased twigs should be pruned atleast 3 to 5 cm below the visible symptoms of the disease. In all cases, the pruned ends should be immediately coated with Chaubattia paint (4 parts of copper carbonate, 4 parts of red-lead and 5 parts of linseed oil.

- Application of fertiliser should be delayed atleast 10 days after pruning. Spraying with copper oxychloride 0.2 per cent or difolatan 0.2 per cent or chlorothalonil 0.2 per cent or rnancozeb 0.2 per cent once in early September and again in late October is recommended for the control of this disease.

- The varieties which are resistant include Bhim, Blue Moon, Red Gold, Quebec, Summer Queen, Red, Ressolute, Samba, Velhiteen Sign and Whitten Sign. Agnasius, Christian Doir, Confidence, Crimson Glory, Fantal Blue, Faryentee, Kiss of Fire, Pascali, Royal Ascot, Vienna Charm and White Chritmas.

Black Spot: Diplocarpon Rosae

Symptoms: The disease is characterized by the presence of black spot on the leaves. These spots are more or less circular in outline. They have a very irregular fibrillose border due to the radiating strands of mycelium which occur beneath the leaves and leaf buds which open late in the season. The plant blossoms poorly. They may not flowers in the following season. On the stem the infected areas present a blackened, blistered appearance, dotted with ustules.

Favourable conditions: The fungus survives in the infected leaves on the plants. The disease is favoured by high humidity and low temperature (21 °C). Winter frosts favour the disease.

Management

- As the fungus perpetuates on old diseased leaves and stems it is necessary to collect and destroy them at the end of the season.

- Diseased plants should be pruned carefully and should be burnt. Spraying with tridemorph 0.025 per cent or captan 0.2 per cent or or Azoxystrobin 0.1% ferbam or benomyl 0.1 per cent at weekly intervals starting with the sprouting of the plants till the appearance of the new foliage and continuing during humid weather effectively controls the disease. Captan sprays at 15 days interval is effective in the control of black spot.

- Shade and excessive irrigation should be avoided.

- Hybrid Rugosa rose, 'Martin Frobisher1 is immune to black spot. Rose cultivars viz., Belaya (Rosa alba). John Cabot and Carefree Beauty are resistant. Among the HT cultivars Show Girl, Buccaneer, Gold Crown Mc-Gredy's Sunset and Perfecta are less susceptible.

Rust Phragmidium Mucronatum

Symptoms: The under-side of the leaves, stems show orange to lemon yellow pustules (1.0 mm in dia) which increase in size as the season advances. In the mid-summer, the orange yellow spots on the leaves are replaced by brick red spots. Later in the season, the same leaves show minute black, hair-like tufts on the under surface. The affected leaves turn yellow, deformed and fall prematurely to the ground.Some times blossoms develop badly or not at all. The diseased bushes are greatly weakened and may die back.

Favourable conditions: Maximum rust infection occurs at a temperature which ranges from 18 to 21 °C. Temperature between 20 and 25 °C is favourable for uredospore production.

Teleutospores which are produced in autumn helps in overwintering and causing fresh infection through basidiospore in the next spring. The fungus also overwinters as perennial mycelium in the stem. Secondary spread is through wind-borne uredospores.

Management

- Diseased, fallen leaves should be collected and burnt.

- Stems harbouring perennial mycelium should be cut out and burnt.

- Three sprayings during Mar-Apr at 15 days interval with mancozeb 0.2 per cent or carbox-in 0.1 per cent. The disease is controlled by spraying with ferbam 0.2 per cent or wettable sulphur 0.3 per cent or captan 0.2 per cent.

Botrytis Bud and Twig Blight - Botrytis Cinerea

Symptoms: The disease is also known as petal fire or Botrytis mold. Infection starts from the sepals as black-brown specks that cover the flower in due course. The buds turn brown and decay. Sometimes partially opened buds are attacked, and the individual petals turn brown and shrivel. In cool moist weather the flower is covered with greenishgrey or darkish growth of the fungus.

Management

- Picking and destroying old blooms and overwintered canes help in reducing the disease.

- Avoiding excessive irrigation helps to check the disease.

- Fungicidal spray with with ferbam 0,2 per cent or captan 0.2 per centor benomyl 0.1 per cent or mancozeb 6.2 per cent or carbendazim 0.2 percent.

- The following rose varieties viz., Anieval Sparks, Bonnienuit, Chantare, Charleston, Devine, Elizabeth, Glimpses, Golden Giant, Joseph's Coat, Picture, Purna, Rakat Gandha, Sharella, Spartan and Zenium Mukhatis are free from the disease. Buds and twig blight by Phomopsis gulabia from Uttar Pradesh has also been reported.

Diseases of Jasmine

- Leaf spots: Alternaria Leaf Blight : Alternaria jasmine

- Alternaria Leaf Spot: Alternaria Alternate

- Cercospora Leaf Spot: Cercospora Jasminicola

Symptoms: On the affected leaves dark brown spots in concentric rings or circular to irregular reddish brown spots of 2 to 8 mm dia appear on upper leaf surface are noticed. During humid conditions, the spots in each leaflet enlarge very quickly and coalesce. Later, blighted leaflets dry and easily fall off. They are also brittle. In a severely affected, garden, large number of fallen-leaves can be easily seen on the ground near the base of the diseased vine. Oval to elongated light brown spots develop on petioles, stem, calyx and even on tubular corollas. In severe cases of infection vegetative buds and young branches dry up. The disease spreads through wind-borne conidia.

Favourable conditions: The disease is severe during winter months (Oct - Dec). In certain areas the disease is noticed even upto February.

Management

Diseased and fallen leaves should be collected and burnt. Spraying mancozeb 0.2 % or Azoxystrobin 0.1% Spraying can be repeated at 7 to 10 days interval covering all the foliage in the vines.

Collar Rot and Root Rot – Sclerotium Rolfsii

Symptoms: Plants at all stages are infected. First the older leaves become yellow followed by younger leaves and finally death of the plant. In the root black discoloration can be seen. On the infected tissues and stem surface white strands of mycelia and mustard like sclerotia are seen.

Management

- Soil drenching with Trifloxystrobin + Tebuconazole @ 0.75 g/lit or Difenoconazole @ 0.5 ml/lit.

- Soil application of Pseudomonas fluorescens @ 25 g/m² and foliar application of P. Fluorescens @ 5 g/lit at monthly intervals after planting. Soil drenching with Copper oxychloride 0.25% or 1% Bordeaux mixture or application of FYM with Trichoderma viride @ 10 g + 100g FYM/plant.

Phyllody: Phytoplasma

- Symptoms: Leaves become small malformed and bushy. In the place of flowers green leaf like malformed flowers are formed. The disease is transmitted by grafting and whitefly, Dialeurodes kirkaldii.

- Management: Selection of cuttings from healthy plants. Spraying insecticide to control the vector.

- Powdery mildew: Oidium jasmini.

- Symptoms: The disease appears as white powdery patches on the upper surface of the leaves. Later, these patches coalesce and cover the entire surface and blight the leaves.

- Management: Spray with Weatble suplur 0.2 % or Chlorothlonil 0.1% or Azoxystrobin @ 1ml/lit.

Diseases of Chrysanthemum

- Blotch/Leaf spot: Septoria chrysanthemella.

- Symptoms: Blackish-brown, circular to irregular spots surrounded by an yellow halo. They coalesce with one another and form large patches coalesce to form blotches covering major portion of the leaf. The dead leaves hang on the stem for sometime.

- Favourable conditions: Infected debris in the soil appeared to be the primary source of infection. The disease is particularly severe during and after monsoon and is favoured by cool weather. Warm weather is not conducive for its development.

- Management: Diseased plant debris should be collected and burnt. Irrigation should be regulated. Fortnightly spraying with carbendazim 0.1 per cent or benomyl 0.1 per cent or mancozeb 0.2 per cent or copper oxychloride 0.3 per cent or 0.1% azoxystrobin, or 0.2% chlorothalonil, thiophanate methyl @ 0.1%. Chrysanthemum cultivars *viz.*, Alpana, Aparjito, C.L. Philips, Flirt, Liliput, Phillies and Sarad are highly resistant.

White Rusts: Puccinia Horiana

Symptoms: Infection by first noticed as yellow to tan spots on the upper surface of the leaves, up to 5 mm diameter; the centers of the spots later turn brown. On the underside of the leaves raised buff, pinkish, waxy pustules develop which later become whitish and quite. Severe infections can lead to complete loss of the crop.

Control

Regulations require that infect plants be destroyed to prevent disease establishment in this country. Protect healthy plants with fungicides with the active ingredients 0.1% azoxystrobin, or Difenconazole 0.05 % or chlorothalonil 0.2%, or thiophanate methyl @ 0.1%.

Vascular Wilts

Chrysanthemums are subject to two vascular wilt diseases caused by Fusarium oxysporum f.sp. chrysanthemi and Verticillium dahliae. Both pathogens persist in the soil for many years.

- Fusarium Wilt: Yellowing of foliage, stunting and wilting often along one side of plant. Plants may appear water stressed and foliage may brown and die. Stems - reddish brown discoloration of the vascular system spread in contaminated soil and infected cuttings and is favored by warm temperatures.

- Verticillium Wilt: Symptoms of Verticillium wilt often appear only after blossom buds have formed; young vigorous plants may be symptomless. Foliage becomes yellow and wilted, sometimes only along leaf margins and on one side of the plant. Leaves begin to die from the base of the plant upward and often remain attached. Stems exhibit dark streaks in the vascular system. Favored when cool weather is followed by hot temperatues. pasteurized growing media and pathogen-free cuttings. Most cultivars are resistant. Avoid susceptible cultivars including 'Bright Golden Ann', 'Echo', 'Glowing Mandalay', Mountain Peak', 'Paragon', 'Pert', 'Puritan', and 'Wedgewood', high relative humidity, overwatering, and poor drainage.

- Management: Pathogen free cuttings or plants and pasteurized growing media. Adjust pH to 6.5 to 7.0 and use nitrate nitrogen fertilization. Soil drenching with Copper oxychloride 2.5 g/lit or Trifloxystrobin + Tebuconazole @0.75 g/lit or Difenoconazole @0.5ml/lit. Avoid highly susceptible cultivars such as 'Bravo', 'Cirbronze', 'Illini Trophy', 'Orange Bowl', 'Royal Trophy', and 'Yellow Delaware'.

Powdery Mildew: Oidium Chrysanthemi

Symptoms: The leaves get covered with a whitish, ash-grey powdery growth on the upper surface. Infected leaves turn yellow and dry. Severely infected plants remain stunted and do not flower. The disease is favoured by dry hot weather. Shade and overcrowding of plants should be avoided to reduce the disease. Spraying with wettable sulphur 0.2 per cent or triforine 0.03 per cent or thiophanate-methyl 0.05 per cent or dinocap 0.025 per cent or dinocap 1.0 kg/ha or cabendazim or benomyl 0.1 per cent at 10 to 15 days interval controls the disease.

Diseases of Crossandra

Wilt: Fusarium Solani

- Symptoms: Disease is observed one month after transplanting. Leaves of infected plants become pale and droop. Margin of the leaves show pinkish brown discolouration. The discolouration spreads to the midrib in a period of 7 to 10 days. Stem portion gets shrivelled. Dark lesions are noticed on the roots extending upto collar region which result in sloughing off the cortical tissue.

- Fvaourable conditions: The disease is formed in both air black and sandy loam soil and losses upto 80 per cent of plants has been reported. Pathogen survive in soil and they are spread by irrigation water. Incidence is more in the presence of root lesion nematode, Pratylenchus delatrei and Helicohylenchus dihystera.

- Management: Affected plants should be pulled out and destroyed to reduce the disease. The nematode can be controlled by soil application of phorate at the rate of 1g/plant on 10th day of transplanting. Soil drenching with carbendazim 0.1 per cent or copper oxychloride 0.25 per cent on 30 days

- Alternaria leaf spot: Alternaria amaranthi var. crossandrae Symptoms. This disease was first reported from Tamil Nadu during 1972. Infected leaves show small, circular or irregular yellow spots on the upper surface.

- They soon enlarge turn brown and develop dark brown concentric rings. Infected leaves become yellow and drop off prematurely.

- Management: Spraying with Mancozeb 0.2% (or) Carbendazim 0.1%.

Leaf Blight: Colletotrichum Crossandra

- Symptoms: The symptoms of leaves consist of the development of brownish, depressed necrotic areas surrounded by reddish and slightly raised margins. Initially the spots appear as brownish specks but become darker as they expand. The lesions are more prominent on lower leaves and confined to the margins.Infected leaves roll up, shrivel and drop off, leaving a barren stem with a whorl of young leaves at the top.

- Management: Spraying with Mancozeb 0.2% (or) Carbendazim 0.1%.

Diseases of Marigold

Rust: Uromyces Dianthi

- Symptoms: On leaves, stems, or flower buds are small, slightly raised blisters that eventually rupture, forming pustules filled with powdery reddish-brown spores. A yellow margin surrounds the pustules and, when infections are severe, entire leaves turn yellow and die. Stems may be girdled when several pustules develop around the shoot, resulting in decreased flower production and quality. Plants may be attacked at any stage of development.

- Favorable conditions: Survives in infected cuttings. Disease is favored by cool nights alternating with warm humid days. This induces dew at night and the formation of a film of water on the leaf surface. More severe in open air culture and in plastic film greenhouses, where dew formation is common.

- Management: Spraying with 0.1% or 0.25 % chlorothalonil, or thiophanate methyl @ 0.1%.

Alternaria leaf Blight- Alternaia Dianthi, Flower Blight - A.Diathicola

- Alternaia dianthi: Tiny purple spots enlarge, in to large lesions with a purple margin and a yellow-green border surrounding a gray-brown center covered with black spores Several lesions may expand and coalesce to form large, irregular necrotic areas that eventually kill the entire leaf. The branches are most frequently infected at the nodes and branch base. These infection centers enlarge to form cankers, which eventually girdle the stem, causing the branch to wilt and the girdled portion to turn yellow and die.

- Adiathicola: Tan to dark brown lesions on sepals and petals . These lesions are covered with dark brown powdery spores that are disseminated by wind and rain. Extended wet periods with light night rains favor outbreaks of this disease.

- Management: Spray with Mancozeb 0.2% or Copper oxy chloride@ 0.2% or Propiconazole 0.1%.

Diseases of Tuberose

- Blight/Leaf and flower spot: Botrytis elliptica.

- Spots are orange to reddish brown and oval on the leaves. They coalesce and blight the leaf. Infection starts from the lower leaves. If the disease occurs early, the entire apical growth of the plant is killed. Flower buds rot or open to distorted flowers with irregular brown flecks.

- Favourable conditions: The disease spreads by rain, air currents and gardeners. The fungussurvives as sclerotia in fallen flowers and leaves. Optimum temperature for spore germination is 15.6 °C. The disease development is favoured at 21 °C.

- Management: Dense planting, shady or low spots with little air circulation should be avoided. The diseased plant parts should be removed .Spraying with Bordeaux mixture 1.0 per cent for every 15 days controls the disease.

Diseases of Carnation

- Fusarium wilt: Fusarium oxysporum f. sp. dianthi.

- Symptoms: In young plants, the first sign of the disease is fading or greying of the normal colour of the leaves with wilting of the leaves and young stems. It is followed by eventual collapse of the whole plant. When older plants are infected, similar symptoms are produced but the older leaves may show chlorosis followed by an indistinct purple-red discolouration. The vascular tissues of infected stems is stained dark brown. Mature plants show wilt symptoms over a period of several months before they die and eventually become straw coloured.

- Favorable conditions: The pathogen is soil-borne and survives and spreads through irrigation water.Warm temperature favours the disease.

- Management: The diseased plants should be removed immediately after noticing the disease. Soil drenching with Carbendazim @ 1 g/lit or Difenoconazole @ 0.5 ml/lit at weekly intervals.Pseudomonas fluorescens as soil application @ 15 g/m² and foliar application @ 5 g/lit at monthly intervals.Soil drenching with Bacillus amyloliquefaciens @ 5 ml/lit at-monthly intervals Grafting of susceptible cultivars like Alice, Fulvio Rosa, Gus Royalette and Johy. on to resistant rootstocks *i.e.* Arancio 25D, Exquisite, Heidi and May Britt and growing in soil naturally infested with fungus was also found to reduce the incidence of disease.

Alternaria Leaf Spot: Alternaria Dianthi

- Symptoms: The chief symptom is blight or rot at leaf bases and around nodes, which are girdled. Spots on leaves are ashy white. The centre of old spots are covered with dark brown to black fungal growth. Leaves may be constricted and twisted and the tip may be killed. Branches die- back at the girdled area and black crusts of conidia are formed on the cankers.

- Favorable conditions: Conidia are spread during watering or in rains. The conidia are carried by on cuttings. The disease is widespread in humid weather.

- Management: To reduce the disease incidence, humidity may be kept low by providing proper air circulation. Disease-free planting material should be used. Spray Tebuconazole @ 2 ml/lit or Propiconazole @ 2 ml/lit. Bacillus subtilis as soil application @ 15 g/m² followed by foliarapplication @ 5 g/lit at monthly intervals.

Cottony Rot: Sclerotinia Sclerotiorum

- Symptoms: Stems rotted; flower rot is similar to gray mold. Cottony, white fungal mass may occur on rotted tissues. Black sclerotia may form inside or outside the stem.

- Favorable conditions Survives in soil and in infected plant debris.

- Favored by high humidity.

- Management: Spray foliage with iprodione or thiophanate-methyl@ 0.1%.

References

- Floriculture, agriculture: vikaspedia.in, Retrieved 10 March, 2019

- Importance-and-scope-of-commercial-floriculture: krishijagran.com, Retrieved 21 May, 2019

- 8-main-steps-for-growing-roses-horticulture, roses, horticulture: biologydiscussion.com, Retrieved 3 February, 2019

- Chrysanthemum, horticulture: indiaagronet.com, Retrieved 17 August, 2019

- Growing-chrysanthemum: gilmour.com, Retrieved 7 May, 2019

- Liu, P. L., et al. (2012). Phylogeny of the genus Chrysanthemum L.: Evidence from single-copy nuclear gene and chloroplast DNA sequences. PLoS ONE 7(11), e48970. doi:10.1371/journal.pone.0048970

- Chrysanthemum-cultivation: agrifarming.in, Retrieved 20 January, 2019

- Methods-of-propagation-chrysanthemum, floriculture: biologydiscussion.com, Retrieved 21 April, 2019

- Chrysanthemum-cultivation: agrifarming.in, Retrieved 19 July, 2019

- Hyam, R. & Pankhurst, R.J. (1995). Plants and their names : a concise dictionary. Oxford: Oxford University Press. p. 186. ISBN 978-0-19-866189-4

- Marigolds, flowersandseasons, growingflowers: theflowerexpert.com, Retrieved 9 May, 2019

- Marigold-varieties-soil-and-climate-horticulture, floriculture: biologydiscussion.com, Retrieved 28 February, 2019

- Tuberose, flowers, crop-production, agriculture: vikaspedia.in, Retrieved 30 April, 2019

- Jefferson-Brown, Michael (2008). Lilies (Wisley handbooks). United Kingdom: Mitchell Beazley. p. 96. ISBN 978-1-84533-384-3

- Tuberose-harvesting-and-diseases-flowers, floriculture: biologydiscussion.com, Retrieved 12 June, 2019

Chapter 5

Organic Horticulture

The science and practice of growing fruits, vegetables, flowers and ornamental plants by following the essential principles of organic agriculture is termed as organic horticulture. It consists of a holistic and sustainable approach towards horticulture, often using natural processes which take place over an extended period of time. All the diverse principles of organic horticulture have been carefully analyzed in this chapter.

Organic Farming

Organic farming is a production system which avoids or largely excludes the use of synthetically compounded fertilizers, pesticides, growth regulators, genetically modified organisms and livestock food additives. To the maximum extent possible organic farming system rely upon crop rotations, use of crop residues, animal manures, legumes, green manures, off farm organic wastes, biofertilizers, mechanical cultivation, mineral bearing rocks and aspects of biological control to maintain soil productivity and tilth to supply plant nutrients and to control insect, weeds and other pests.

Organic methods can increase farm productivity, repair decades of environmental damage and knit small farm families into more sustainable distribution networks leading to improved food security if they organize themselves in production, certification and marketing. During last few years an increasing number of farmers have shown lack of interest in farming and the people who used to cultivate are migrating to other areas. Organic farming is one way to promote either self-sufficiency or food security. Use of massive inputs of chemical fertilizers and toxic pesticides poisons the land and water heavily. The after-effects of this are severe environmental consequences, including loss of topsoil, decrease in soil fertility, surface and ground water contamination and loss of genetic diversity.

Organic farming which is a holistic production management system that promotes and enhances agro-ecosystem health, including biodiversity, biological cycles, and soil biological activity is hence important. Many studies have shown that organic farming methods can produce even higher yields than conventional methods. Significant difference in soil health indicators such as nitrogen mineralization potential and microbial abundance and diversity, which were higher in the organic farms can also be seen. The increased soil health in organic farms also resulted in considerably lower insect and disease incidence. The emphasis on small-scale integrated farming systems has the potential to revitalize rural areas and their economies.

Reasons for Organic Farming

The population of the planet is skyrocketing and providing food for the world is becoming extremely difficult. The need of the hour is sustainable cultivation and production of food for all. The Green

Revolution and its chemical based technology are losing its appeal as dividends are falling and returns are unsustainable. Pollution and climate change are other negative externalities caused by use of fossil fuel based chemicals.

In spite of our diet choices, organic food is the best choice you'll ever make, and this means embracing organic farming methods. Here are the reasons why we need to take up organic farming methods:

1. To accrue the benefits of nutrients: Foods from organic farms are loaded with nutrients such as vitamins, enzymes, minerals and other micro-nutrients compared to those from conventional farms. This is because organic farms are managed and nourished using sustainable practices. In fact, some past researchers collected and tested vegetables, fruits, and grains from both organic farms and conventional farms.

The conclusion was that food items from organic farms had way more nutrients than those sourced from commercial or conventional farms. The study went further to substantiate that five servings of these fruits and vegetables from organic farms offered sufficient allowance of vitamin C. However, the same quantity of fruits and vegetable did not offer the same sufficient allowance.

2. Stay away from GMOs: Statistics show that genetically modified foods (GMOs) are contaminating natural foods sources at real scary pace, manifesting grave effects beyond our comprehension. What makes them a great threat is they are not even labeled. So, sticking to organic foods sourced from veritable sources is the only way to mitigate these grave effects of GMOs.

3. Natural and better taste: Those that have tasted organically farmed foods would attest to the fact that they have a natural and better taste. The natural and superior taste stems from the well balanced and nourished soil. Organic farmers always prioritize quality over quantity.

4. Direct support to farming: Purchasing foods items from organic farmers is a surefire investment in a cost-effective future. Conventional farming methods have enjoyed great subsidies and tax cuts from most governments over the past years. This has led to the proliferation of commercially produced foods that have increased dangerous diseases like cancer. It's time governments invested in organic farming technologies to mitigates these problems and secure the future. It all starts with you buying food items from known organic sources.

5. To conserve agricultural diversity: These days, it normal to hear news about extinct species and this should be a major concern. In the last century alone, it is approximated that 75 percent of agricultural diversity of crops has been wiped out. Slanting towards one form of farming is a recipe for disaster in the future. A classic example is a potato. There were different varieties available in the marketplace. Today, only one species of potato dominate.

This is a dangerous situation because if pests knock out the remaining potato specie available today, we will not have potatoes anymore. This is why we need organic farming methods that produce disease and pest resistant crops to guarantee a sustainable future.

6. To prevent antibiotics, drugs, and hormones in animal products: Commercial dairy and meat are highly susceptible to contamination by dangerous substances. A statistic in an American journal revealed that over 90% of chemicals the population consumes emanate from meat tissue and

dairy products. According to a report by Environmental Protection Agency (EPA), a vast majority of pesticides are consumed by the population stem from poultry, meat, eggs, fish and dairy product since animals and birds that produce these products sit on top of the food chain.

This means they are fed foods loaded with chemicals and toxins. Drugs, antibiotics, and growth hormones are also injected into these animals and so, are directly transferred to meat and dairy products. Hormone supplementation fed to farmed fish, beef and dairy products contribute mightily to ingestion of chemicals. These chemicals only come with a lot of complications like genetic problems, cancer risks, growth of tumor and other complications at the outset of puberty.

Key Features of Organic Farming

- Protecting soil quality using organic material and encouraging biological activity.

- Indirect provision of crop nutrients using soil microorganisms.

- Nitrogen fixation in soils using legumes.

- Weed and pest control based on methods like crop rotation, biological diversity, natural predators, organic manures and suitable chemical, thermal and biological intervention.

- Rearing of livestock, taking care of housing, nutrition, health, rearing and breeding.

- Care for the larger environment and conservation of natural habitats and wildlife.

Four Principles of Organic Farming

- Principle of Health: Organic agriculture must contribute to the health and well being of soil, plants, animals, humans and the earth. It is the sustenance of mental, physical, ecological and social well being. For instance, it provides pollution and chemical free, nutritious food items for humans.

- Principle of Fairness: Fairness is evident in maintaining equity and justice of the shared planet both among humans and other living beings. Organic farming provides good quality of life and helps in reducing poverty. Natural resources must be judiciously used and preserved for future generations.

- Principle of Ecological Balance: Organic farming must be modeled on living ecological systems. Organic farming methods must fit the ecological balances and cycles in nature.

- Principle of Care: Organic agriculture should be practiced in a careful and responsible manner to benefit the present and future generations and the environment.

As opposed to modern and conventional agricultural methods, organic farming does not depend on synthetic chemicals. It utilizes natural, biological methods to build up soil fertility such as microbial activity boosting plant nutrition.

Secondly, multiple cropping practiced in organic farming boosts biodiversity which enhances productivity and resilience and contributes to a healthy farming system. Conventional farming systems use mono cropping that destroys the soil fertility.

Unsustainability of Modern Farming

1. Loss of soil fertility due to excessive use of chemical fertilizers and lack of crop rotation.

2. Nitrate run off during rains contaminates water resources.

3. Soil erosion due to deep ploughing and heavy rains.

4. More requirement of fuel for cultivation.

5. Use of poisonous bio-cide sprays to curb pest and weeds.

6. Cruelty to animals in their housing, feeding, breeding and slaughtering.

7. Loss of biodiversity due to mono culture.

8. Native animals and plants lose space to exotic species and hybrids.

Benefits of Organically Grown Food Items and Agricultural Produce

-   Better Nutrition: As compared to a longer time conventionally grown food, organic food is much richer in nutrients. Nutritional value of a food item is determined by its mineral and vitamin content. Organic farming enhances the nutrients of the soil which is passed on to the plants and animals.

-   Helps us stay healthy: Organic foods do not contain any chemical. This is because organic farmers don't use chemicals at any stage of the food-growing process like their commercial counterparts. Organic farmers use natural farming techniques that don't harm humans and environment. These foods keep dangerous diseases like cancer and diabetes at bay.

-   Free of poison: Organic farming does not make use of poisonous chemicals, pesticides and weedicides. Studies reveal that a large section of the population fed on toxic substances used in conventional agriculture have fallen prey to diseases like cancer. As organic farming avoids these toxins, it reduces the sickness and diseases due to them.

-   Organic foods are highly authenticated: For any produce to qualify as organic food, it must undergo quality checks and the creation process rigorously investigated. The same rule applies to international markets. This is a great victory for consumers because they are getting the real organic foods. These quality checks and investigations weed out quacks who want to benefit from the organic food label by delivering commercially produced foods instead.

- Lower prices: There is a big misconception that organic foods are relatively expensive. The truth is they are actually cheaper because they don't require application of expensive pesticides, insecticides, and weedicides. In fact, you can get organic foods direct from the source at really reasonable prices.

- Enhanced Taste: The quality of food is also determined by its taste. Organic food often tastes better than other food. The sugar content in organically grown fruits and vegetables provides them with extra taste. The quality of fruits and vegetables can be measured using Brix analysis.

- Organic farming methods are eco-friendly: In commercial farms, the chemicals applied infiltrate into the soil and severely contaminate it and nearby water sources. Plant life, animals, and humans are all impacted by this phenomenon. Organic farming does not utilize these harsh chemicals so; the environment remains protected.

- Longer shelf–life: Organic plants have greater metabolic and structural integrity in their cellular structure than conventional crops. This enables storage of organic food for a longer time.

Organic farming is preferred as it battles pests and weeds in a non-toxic manner, involves less input costs for cultivation and preserves the ecological balance while promoting biological diversity and protection of the environment.

Nutrient Management in Organic Farming

In organic farming, it is important to constantly work to build a healthy soil that is rich in organic matter and has all the nutrients that the plants need. Several methods viz. green manuring, addition of manures and biofertilizers etc can be used to build up soil fertility. These organic sources not only add different nutrients to the soil but also help to prevent weeds and increase soil organic matter to feed soil microorganisms. Soil with high organic matter resists soil erosion, holds water better and thus requires less irrigation. Some natural minerals that are needed by the plants to grow and to improve the soil's consistency can also be added. Soil amendments like lime are added to adjust the soil's pH balance. However soil amendment and water should contain minimum heavy metals. Most of the organic fertilizers used are recycled by-products from other industries that would otherwise go to waste. Farmers also make compost from animal manures and mushroom compost. Before compost can be applied to the fields, it is heated and aged for at least two months, reaching and maintaining an internal temperature of 130°-140 °F to kill unwanted bacteria and weed seeds. A number of organic fertilizers/amendments and bacterial and fungal biofertilizers can be used in organic farming depending upon availability and their suitability to crop. Different available organic inputs are described below.

Organic Manures

Commonly available and applied farm yard manure (FYM) and vermicompost etc. are generally low in nutrient content, so high application rates are needed to meet crop nutrient requirements. However, in many developing countries the availability of organic manures is not sufficient for crop requirements; partly due to its extensive use of cattle dung in energy production. Green manuring with Sesbania, cowpea, green gram etc are quiet effective to improve the organic matter content of soil. However, use of green manuring has declined in last few decades due to intensive cropping and socioeconomic reasons. Considering these constraints International Federation of Organic Agriculture Movement (IFOAM) and Codex Alimentarius have approved the use of some inorganic sources of plant nutrients like rock phosphate, basic slag, rock potash etc. in organic farming systems. These substances can supply essential nutrients and may be from plant, animal, microbial or mineral origin and may undergo physical, enzymatic or microbial processes and their use does not result in unacceptable effects on produce and the environment including soil organisms.

Bacterial and fungal Biofertilizers

Contribution of biological fixation of nitrogen on surface of earth is the highest (67.3%) among all the sources of N fixation. Following bacterial and fungal biofertilizers can be used as a component of organic farming in different crops.

- Rhizobium: The effectiveness of symbiotic N_2 fixing bacteria viz. Rhizobia for legume crops eg. Rhizobium, Bradyrhizobium, Sinorhizobium, Azorhizobium, and Mesorhizobium etc have been well recognized. These bacteria infecting legumes have a global distribution. These rhizobia have a N_2-fixing capability up to 450 kg N ha−1 depending on host-plant species and bacterial strains. Carrier based inoculants can be coated on seeds for the introduction of bacterial strains into soil.

- Azotobacter: N_2 fixing free-living bacteria can fix atmospheric nitrogen in cereal crops without any symbiosis. Such free living bacterias are: Azotobacter sp. for different cereal crops; Acetobacter diazotrophicus and Herbaspirillum spp. for sugarcane, sorghum and maize crop. Beside fixing nitrogen, they also increase germination and vigour in young plants leading to an improved crop stand. They can fix 15-20 kg/ha nitrogen per year. Azotobacter sp. also has ability to produce anti fungal compounds against many plant pathogens. Azotobacter can biologically control the nematode diseases of plants also.

- Azospirillum: The genus Azospirillum colonizes in a variety of annual and perennial plants. Studies indicate that Azospirillum can increase the growth of crops like sunflower, carrot, oak, sugarbeet, tomato, pepper, cotton, wheat and rice. The crop yield can increase from 5-30%. Inoculum of Azotobacter and Azospirillum can be produced and applied as in peat formulation through seed coating. The peat formulation can also be directly utilized in field applications.

- Plant growth promoting rhizobacteria: Various bacteria that promote plant growth are collectively called plant growth promoting rhizobacteria (PGPR). PGPR are thought to improve plant growth by colonizing the root system and pre empting the establishment of suppressing deleterious rhizosphere microorganisms on the roots. Large populations of bacteria established in planting material and roots become a partial sink for nutrients in the rhizosphere thus reducing the amount of C and N available to stimulate spores of fungal pathogens or for subsequent colonization of the root. PGPR belong to several genera viz.Actinoplanes, Azotobacter, Bacillus, Pseudomonas, Rhizobium, Bradyrhizobium, Streptomyces, Xanthomonas etc. Bacillus spp. act as biocontrol agent because their endospores are tolerant to heat and desiccation. Seed treatment with B.subtilis is reported to increase yield of carrot by 48%, oats by 33% and groundnut upto 37%.

- Phosphorus-solubilizing bacteria (PSB): Phosphorus is the vital nutrient next to nitrogen for plants and microorganisms. This element is necessary for the nodulation by Rhizobium and even to nitrogen fixers, Azollaand BGA. The phospho microorganism mainly bacteria and fungi make available insoluble phosphorus to the plants. It can increase crop yield up to 200-500 kg/ha and thus 30 to 50 kg Super Phosphate can be saved. Most predominant phosphorus-solubilizing bacteria (PSB) belong to the genera Bacillus and Pseudomonas. At present PSB is most widely used biofertilizer in India. PSB can reduce the P requirement of crop up to 25%.

- Mycorrhizal fungi: Root-colonizing mycorrhizal fungi increase tolerance of heavy metal contamination and drought. Mycorrhizal fungi improve soil quality also by having a direct influence on soil aggregation and therefore aeration and water dynamics. An interesting potential of this fungi is its ability to allow plant access to nutrient sources which are generally unavailable to the host plants and thus plants may be able to use insoluble sources of P when inoculated with mycorrhizal fungi but not in the absence of inoculation.

- Blue green algae (BGA): BGA are the pioneer colonizers both in hydrosphere and xerosphere. These organisms have been found to synthesize 0.8 x 1011 tonnes of organic matter, constituting about 40 percent of the total organic matter synthesized annually on this planet. BGA constitute the largest, most diverse and widely distributed group of prokaryotic microscopic organisms that perform oxygenic photosynthesis. These are also known as cyanophyceae and cyanobacteria. These are widely distributed in tropics; and are able to withstand extremes of temperature and drought. The significance of the abundance of BGA in rice soils has been well recognized. Multi-location trials conducted under varying agro climatic conditions have indicated that the algal inoculation could save 30 kg N/ha, however, it depends upon the agro ecological conditions. BGA has been reported to reduce the pH of soil and improve upon exchangeable calcium and water holding capacity. The recommended method of application of the algal inoculum is broadcasting on standing water about 3 to 4 days after transplantation. After the application of algal inoculum the field should be kept water logged for about a week's time. Establishment of the algal inoculum can be observed within a week of inoculation in the form of floating algal mats, more prominently seen in the afternoon.

- Azolla: A floating water fern 'Azolla' hosts nitrogen fixing BGA Anabaena azollae. Azolla contains 3.4% nitrogen (on dry wt. basis) and add organic matter in soil. This biofertilizer is used for rice cultivation. There are six species of Azolla viz. A. caroliniana, A. nilotica, A. mexicana, A.filiculoides, A. microphylla and A. pinnata. Azolla plant has a floating, branched stem, deeply bilobed leaves and true roots which penetrate the body of water .The leaves are arranged alternately on the stem. Each leaf has a dorsal and ventral lobe. The dorsal fleshy lobe is exposed to air and contains chlorophyll. It grows well in ditches and stagnant water. Azolla can be easily grown throughout the year if water is not a limiting factor and climatic conditions are favourable for its growth. This fern usually forms a green mat over water. Azolla is readily decomposed to NH_4 which is available to the rice plants. Field trial have shown that rice yields increased by 0.5-2t/ha due to Azolla application.

Weed Management in Organic Farming

In organic farming systems, the aim is not necessarily the elimination of weeds but their control. Weed control means reducing the effects of weeds on crop growth and yield.

Organic farming avoids the use of herbicides which, like pesticides, leave harmful residues in the environment. Beneficial plant life such as host plants for useful insects may also be destroyed by herbicides.

On an organic farm, weeds are controlled using a number of methods:

*   Crop rotation.

*   Hoeing.

*   Mulches, which cover the soil and stop weed seeds from germinating.

*   Hand-weeding or the use of mechanical weeders.

*   Planting crops close together within each bed, to prevent space for weeds to emerge.

*   Green manures or cover crops to outcompete weeds.

*   Soil cultivation carried out at repeated intervals and at the appropriate time, when the soil is moist. Care should be taken that cultivation does not cause soil erosion.

*   Animals as weeders to graze on weeds.

Weeds do have some useful purposes. They can provide protection from erosion, food for animals and beneficial insects and food for human use.

Natural Pest and Disease Control

Pests and diseases are part of nature. In the ideal system there is a natural balance between predators and pests. If the system is imbalanced then one population can become dominant because it is not being preyed upon by another. The aim of natural control is to restore a natural balance between pest and predator and to keep pests and diseases down to an acceptable level. The aim is not to eradicate them altogether.

Chemical Control

Pesticides do not solve the pest problem. In the past 50 years, insecticide use has increased tenfold, while crop losses from pest damage have doubled. Here are three important reasons why natural control is preferable to pesticide use.

Safety for People

Artificial pesticides can quickly find their way into food chains and water courses. This can create health hazards for humans.

Human health can also be harmed by people eating foods (especially fruit and vegetables) which still contain residues of pesticides that were sprayed on the crop.

There is also much concern for those people using chemical pesticides. The products may be misused because the instructions are not written in the language spoken by the person using them. This has led to many accidents such as reports of people suffering from severe skin rashes and headaches as a result of using chemical pesticides. There are an estimated one million cases of poisoning by pesticides each year around the world. Up to 20,000 of these result in death. Most of the deaths occur in tropical countries where chemical pesticides which are banned in Europe or the USA are still available.

Cost

Using natural pest and disease control is often cheaper than applying chemical pesticides because natural methods do not involve buying materials from the outside. Products and materials which are already in the home and around the farm are most often used.

Safety for the Environment

There are a number of harmful effects that chemical pesticides can have on the environment:

- Chemical pesticides can kill useful insects which eat pests. Just one spray can upset the balance between pests and the useful predators which eat them.

- Artificial chemicals can stay in the environment and in the bodies of animals causing problems for many years.

- Insect pests can very quickly, over a few breeding cycles, become resistant to artificial products and are no longer controlled. This means that increased amounts or stronger chemicals are then needed creating further economic, health and environmental problems.

Natural Control

There are many ways in which the organic farmer can control pests and diseases:

- Growing healthy crops that suffer less damage from pests and diseases.

- Choosing crops with a natural resistance to specific pests and diseases. Local varieties are better at resisting local pest and diseases than introduced varieties.

- Timely planting of crops to avoid the period when a pest does most damage.

- Companion planting with other crops that pests will avoid, such as onion or garlic.

Companion Planting

- Trapping or picking pests from the crop.

- Identifying pest and diseases correctly. This will prevent the farmer from wasting time or accidentally eliminating beneficial insects. It is therefore useful to know life cycles, breeding habits, preferred host plants and predators of pests.

- Using crop rotations to help break pest cycles and prevent a carry over of pests to the next season.

- Providing natural habitats to encourage natural predators that control pests. To do this, the farmer should learn to recognise insects and other animals that eat and control pests.

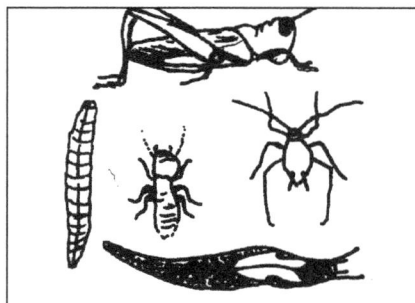

Grasshoppers, slugs, termites, aphids and types of caterpillars are pests.

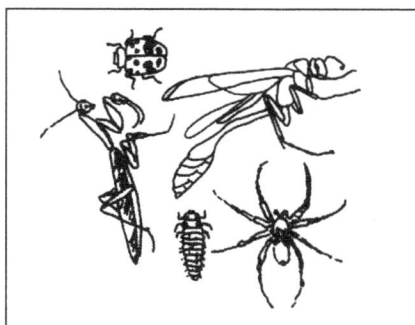

Ladybirds, spiders, ground beetles, parasitic wasps and praying mantis are predators.

Through careful planning and using all the other techniques available it should be possible to avoid the need for any crop spraying. If pests are still a problem natural products can be used to manage pests, including sprays made from chillies, onions, garlic or neem.

Even with these natural pesticides, their use should be limited as much as possible and only the safest ones used. It is wise to check with national and international organic standards to see which ones are allowed or recommended.

Genetic Diversity

Within a single crop there can be many differences between plants. They may vary in height or ability to resist diseases, for example. These differences are genetic.

Traditional crops grown by farmers contain greater genetic diversity than modern bred crops. Traditional varieties have been selected over many centuries to meet the requirements of farmers. Although many are being replaced by modern varieties, seeds are often still saved locally.

Crops which have been bred by modern breeding methods tend to be very similar and if one plant is prone to disease, all the other plants are as well. Although some modern varieties may be very resistant to specific pests and diseases they are often less suited to local conditions than traditional varieties. It can therefore be dangerous to rely too much on any one of them.

Strip cropping onions and tomatoes to prevent pest and disease attack.

In organic systems, some variation or 'genetic diversity' between the plants within a crop is beneficial. Growing a number of different crops rather than relying on one is also very important. This helps to protect against pests and diseases and acts as insurance against crop failure in unusual weather such as drought or flood. It is important to remember this when choosing which crops to grow.

An organic farmer should try to:

- Grow a mixture of crops in the same field (mixed cropping, intercropping, strip cropping).

- Grow different varieties of the same crop.

- Use as many local crop varieties as possible.

save the seed of local and improved crop varieties rather than relying on buying seed from outside the farm every year. Exchange of seed with other farmers can also help to increase diversity, and ensure the survival of the many traditional crop varieties which are being lost as they are replaced by a few modern varieties.

Careful use of Water

In arid lands the careful use of water is as much a part of organic growing as is any other technique. As with other resources, organic farmers should try to use water which is available locally, avoiding using water faster than it is replaced naturally.

There are many ways to use water carefully, including:

1. The use of terracing, rain water basins or catchments and careful irrigation.

2. The addition of organic matter to the soil to improve its ability to hold water.

3. The use of mulches to hold water in the soil by stopping the soil surface from drying out or becoming too hot.

Animal Husbandry

In an organic system, the welfare of the animals is considered very important:

1. Animals should not be kept in confined spaces where they cannot carry out their natural behaviour such as standing and moving around in an inadequate amount of space. However, care should be taken that animals do not damage crops.

2. Food for animals should be grown organically.

3. Breeds should be chosen to suit local needs and local conditions and resources.

These factors help to ensure that livestock are more healthy, better able to resist diseases and to provide good yields for the farmer.

Advantages of Organic farming

1. It helps to maintain environment health by reducing the level of pollution.

2. It reduces human and animal health hazards by reducing the level of residues in the product.

3. It helps in keeping agricultural production at a sustainable level.

4. It reduces the cost of agricultural production and also improves the soil health.

5. It ensures optimum utilization of natural resources for short-term benefit and helps in conserving them for future generation.

6. It not only saves energy for both animal and machine, but also reduces risk of crop failure.

7. It improves the soil physical properties such as granulation, good tilth, good aeration, easy root penetration and improves water-holding capacity and reduces erosion.

8. It improves the soil's chemical properties such as supply and retention of soil nutrients, reduces nutrient loss into water bodies and environment and promotes favourable chemical reactions.

Disadvantages

A 2003 to 2005 investigation by the Cranfield University for the Department for Environment, Food and Rural Affairs in the UK found that it is difficult to compare the Global warming potential, acidification and eutrophication emissions but "Organic production often results in increased burdens, from factors such as N leaching and N_2O emissions", even though primary energy use was less for most organic products. N_2O is always the largest global warming potential contributor except in tomatoes. However, "organic tomatoes always incur more burdens (except pesticide use)". Some emissions were lower "per area", but organic farming always required 65 to 200% more field area than non-organic farming. The numbers were highest for bread wheat (200+ % more) and potatoes (160% more).

Environmental Impact and Emissions

Researchers at Oxford University analyzed 71 peer-reviewed studies and observed that organic products are sometimes worse for the environment. Organic milk, cereals, and pork generated higher greenhouse gas emissions per product than conventional ones but organic beef and olives had lower emissions in most studies. Usually organic products required less energy, but more land. Per unit of product, organic produce generates higher nitrogen leaching, nitrous oxide emissions, ammonia emissions, eutrophication, and acidification potential than conventionally grown produce. Other differences were not significant. The researchers concluded that public debate should consider various manners of employing conventional or organic farming, and not merely debate conventional farming as opposed to organic farming. They also sought to find specific solutions to specific circumstances.

Proponents of organic farming have claimed that organic agriculture emphasizes closed nutrient cycles, biodiversity, and effective soil management providing the capacity to mitigate and even reverse the effects of climate change and that organic agriculture can decrease fossil fuel emissions. "The carbon sequestration efficiency of organic systems in temperate climates is almost double (575–700 kg carbon per ha per year – 510–625 lb/ac/an) that of conventional treatment of soils, mainly owing to the use of grass clovers for feed and of cover crops in organic rotations."

Critics of organic farming methods believe that the increased land needed to farm organic food could potentially destroy the rainforests and wipe out many ecosystems.

Nutrient Leaching

According to a 2012 meta-analysis of 71 studies, nitrogen leaching, nitrous oxide emissions, ammonia emissions, eutrophication potential and acidification potential were higher for organic products, although in one study "nitrate leaching was 4.4–5.6 times higher in conventional plots than organic plots". Excess nutrients in lakes, rivers, and groundwater can cause algal blooms, eutrophication, and subsequent dead zones. In addition, nitrates are harmful to aquatic organisms by themselves.

Land use

The Oxford meta-analysis of 71 studies found that organic farming requires 84% more land for an equivalent amount of harvest, mainly due to lack of nutrients but sometimes due to weeds, diseases or pests, lower yielding animals and land required for fertility building crops. While organic farming does not necessarily save land for wildlife habitats and forestry in all cases, the most modern breakthroughs in organic are addressing these issues with success.

Professor Wolfgang Branscheid says that organic animal production is not good for the environment, because organic chicken requires twice as much land as "conventional" chicken and organic pork a quarter more. According to a calculation by Hudson Institute, organic beef requires three times as much land. On the other hand, certain organic methods of animal husbandry have been shown to restore desertified, marginal, and/or otherwise unavailable land to agricultural productivity and wildlife. Or by getting both forage and cash crop production from the same fields simultaneously, reduce net land use.

In England organic farming yields 55% of normal yields. In other regions of the world, organic methods have started producing record yields.

Pesticides

A sign outside of an organic apple orchard in Pateros, Washington reminding orchardists not to spray pesticides on these trees.

In organic farming synthetic pesticides are generally prohibited. A chemical is said to be synthetic if it does not already exist in the natural world. But the organic label goes further and usually prohibit compounds that exist in nature if they are produced by chemical synthesis. So the prohibition is also about the method of production and not only the nature of the compound.

A non-exhaustive list of organic approved pesticides with their median lethal doses:

- Copper(II) sulfate is used as a fungicide and is also used in conventional agriculture (LD_{50} 300 mg/kg). Conventional agriculture has the option to use the less toxic Mancozeb (LD_{50} 4,500 to 11,200 mg/kg).

- Boric acid is used as stomach poison that target insects (LD_{50}: 2660 mg/kg).

- Pyrethrin comes from chemicals extracted from flowers of the genus Pyrethrum (LD_{50} of 370 mg/kg). Its potent toxicity is used to control insects.

- Lime sulfur (aka calcium polysulfide) and sulfur are considered to be allowed, synthetic materials (LD_{50}: 820 mg/kg).

- Rotenone is a powerful insecticide that was used to control insects (LD_{50}: 132 mg/kg). Despite the high toxicity of Rotenone to aquatic life and some links to Parkinson disease the compound is still allowed in organic farming as it is a naturally occurring compound.

- Bromomethane is a gas that is still used in the nurseries of Strawberry organic farming.

- Azadirachtin is a wide spectrum very potent insecticide. Almost non toxic to mammals (LD_{50} in rats is > 3,540 mg/kg) but affects beneficial insects.

Food Quality and Safety

While there may be some differences in the amounts of nutrients and anti-nutrients when organically produced food and conventionally produced food are compared, the variable nature of food production and handling makes it difficult to generalize results, and there is insufficient evidence to make claims that organic food is safer or healthier than conventional food. Claims that organic food tastes better are not supported by evidence.

Soil Conservation

Supporters claim that organically managed soil has a higher quality and higher water retention. This may help increase yields for organic farms in drought years. Organic farming can build up soil organic matter better than conventional no-till farming, which suggests long-term yield benefits from organic farming. An 18-year study of organic methods on nutrient-depleted soil concluded that conventional methods were superior for soil fertility and yield for nutrient-depleted soils in cold-temperate climates, arguing that much of the benefit from organic farming derives from imported materials that could not be regarded as self-sustaining.

In Dirt - The Erosion of Civilizations, geomorphologist David Montgomery outlines a coming crisis from soil erosion. Agriculture relies on roughly one meter of topsoil, and that is being depleted ten times faster than it is being replaced. No-till farming, which some claim depends upon pesticides, is one way to minimize erosion. However, a 2007 study by the USDA's Agricultural Research Service has found that manure applications in tilled organic farming are better at building up the soil than no-till.

Biodiversity

The conservation of natural resources and biodiversity is a core principle of organic production. Three broad management practices (prohibition/reduced use of chemical pesticides and inorganic fertilizers; sympathetic management of non-cropped habitats; and preservation of mixed farming) that are largely intrinsic (but not exclusive) to organic farming are particularly beneficial for farmland wildlife. Using practices that attract or introduce beneficial insects, provide habitat for birds and mammals, and provide conditions that increase soil biotic diversity serve to supply vital ecological services to organic production systems. Advantages to certified organic operations that implement these types of production practices include: 1) decreased dependence on outside fertility inputs; 2) reduced pest management costs; 3) more reliable sources of clean water; and 4) better pollination.

Nearly all non-crop, naturally occurring species observed in comparative farm land practice studies show a preference for organic farming both by abundance and diversity. An average of 30% more species inhabit organic farms. Birds, butterflies, soil microbes, beetles, earthworms, spiders, vegetation, and mammals are particularly affected. Lack of herbicides and pesticides improve biodiversity fitness and population density. Many weed species attract beneficial insects that improve soil qualities and forage on weed pests. Soil-bound organisms often benefit because of increased bacteria populations due to natural fertilizer such as manure, while experiencing reduced intake of herbicides and pesticides. Increased biodiversity, especially from beneficial soil microbes and mycorrhizae have been proposed as an explanation for the high yields experienced by some organic plots, especially in light of the differences seen in a 21-year comparison of organic and control fields.

Biodiversity from organic farming provides capital to humans. Species found in organic farms enhance sustainability by reducing human input (e.g., fertilizers, pesticides).

The USDA's Agricultural Marketing Service (AMS) published a *Federal Register* notice on 15 January 2016, announcing the National Organic Program (NOP) final guidance on Natural Resources and Biodiversity Conservation for Certified Organic Operations. Given the broad scope of natural resources which includes soil, water, wetland, woodland and wildlife, the guidance provides examples of practices that support the underlying conservation principles and demonstrate compliance with USDA organic regulations 205.200. The final guidance provides organic certifiers and farms with examples of production practices that support conservation principles and comply with the USDA organic regulations, which require operations to maintain or improve natural resources. The final guidance also clarifies the role of certified operations (to submit an OSP to a certifier), certifiers (ensure that the OSP describes or lists practices that explain the operator's monitoring plan and practices to support natural resources and biodiversity conservation), and inspectors (onsite inspection) in the implementation and verification of these production practices.

A wide range of organisms benefit from organic farming, but it is unclear whether organic methods confer greater benefits than conventional integrated agri-environmental programs. Organic farming is often presented as a more biodiversity-friendly practice, but the generality of the beneficial effects of organic farming is debated as the effects appear often species- and context-dependent, and current research has highlighted the need to quantify the relative effects of local- and landscape-scale management on farmland biodiversity. There are four key issues when comparing the impacts on biodiversity of organic and conventional farming: (1) It remains unclear whether a holistic whole-farm approach (i.e. organic) provides greater benefits to biodiversity than carefully targeted prescriptions applied to relatively small areas of cropped and/or non-cropped habitats within conventional agriculture (i.e. agri-environment schemes); (2) Many comparative studies encounter methodological problems, limiting their ability to draw quantitative conclusions; (3) Our knowledge of the impacts of organic farming in pastoral and upland agriculture is limited; (4) There remains a pressing need for longitudinal, system-level studies in order to address these issues and to fill in the gaps in our knowledge of the impacts of organic farming, before a full appraisal of its potential role in biodiversity conservation in agroecosystems can be made.

Opposition to Labor Standards

Organic agriculture is often considered to be more socially just and economically sustainable for farmworkers than conventional agriculture. However, there is little social science research or consensus as to whether or not organic agriculture provides better working conditions than conventional agriculture. As many consumers equate organic and sustainable agriculture with small-scale, family-owned organizations it is widely interpreted that buying organic supports better conditions for farmworkers than buying with conventional producers. Organic agriculture is generally more labor-intensive due to its dependence on manual practices for fertilization and pest removal and relies heavily upon hired, non-family farmworkers rather than family members. Although illnesses from synthetic inputs pose less of a risk, hired workers still fall victim to debilitating musculoskeletal disorders associated with agricultural work. The USDA certification requirements outline growing practices and ecological standards but do nothing to codify labor practices. Independent certification initiatives such as the Agricultural Justice Project, Domestic Fair Trade Working Group, and the Food Alliance have attempted to implement farmworker interests but because these initiatives require voluntary participation of organic farms, their standards cannot be widely enforced.Despite the benefit to farmworkers of implementing labor standards, there is little support among the organic community for these social requirements. Many actors of the organic industry believe that enforcing labor standards would be unnecessary, unacceptable, or unviable due to the constraints of the market.

Organic Gardening

Organic gardening is a type of gardening where chemical pesticides or fertilizers are not employed. Although synthetic materials are not used to aid the growth and overall well-being of plants, it does not mean the plants are left alone to take care of themselves. Organic gardeners only use materials derived from living things such as manures and composts to fertilize the plants.

Several natural pest-warding ways are also used to keep the plants healthy. Organic gardening's popularity is increasing steadily as many people opt for organic food grown without chemicals, which has a much lesser impact on the environment. Organic gardening is a cheaper yet more demanding way of growing organic food. An organic gardener must work with the area's ecological system and make sure the earth sustains minimal damage. By avoiding the use of synthetic fertilizers and chemicals, organic gardeners help prevent the possible pollution such chemicals cause in nearby rivers and soil composition.

Factors such as mulches, cover crops, vermicompost, and manures set organic growing apart from other methods. Organic growers might feature different types of pest control, including but not limited to: using beneficial microorganisms, plant selection, use of companion crops to divert pests, use of insect traps, and crop rotation.

Organic gardening is essentially gardening without using synthetic products like fertilizers and pesticides. It involves the use of only natural products to grow plants in your garden. Organic gardening replenishes the natural resources as it uses them. In organic gardening, you consider your plants as part of the larger natural system that begins with the soil and includes water supply, the wildlife; insects and people. Everyone wants the food we serve to our families as well as our environment to be safe and healthy. A good organic gardener strives to ensure that his or her activities are in harmony with the natural ecosystem and aims at minimizing exploitation as well as replenishing all the resources consumed by his or her garden.

Gardeners and people who have come across the word organic gardening probably usually desire to know what it means. Organic gardening is a terminology that simply refers to growing of plants, vegetables, and fruits in the best natural way without the use of pesticides or synthetic chemical fertilizers. Even so, organic gardening is more than simply avoiding the use pesticides and synthetic fertilizers. Gardening organically encompasses supporting the health of the entire gardening system naturally. It means working in harmony with the natural systems including the soil, water supply, people, and even insects with an ultimate aim of minimizing destruction to living and non-living things in the natural environment while constantly replenishing any resources utilized during gardening.

Organic gardening fundamentals requires cultivation emphasized on creating an ecosystem that nourishes and sustains soil microbes and plants, while also benefiting insects rather than just putting seeds in the ground and letting them grow.

To sum up, organic gardening is as simple as relying on intermingling plant types and varieties, use of companion planting, dense planting in order for some plants to offer companion to vulnerable plants, and supporting natural systems to minimize spread of pests and diseases. To begin with, there are three major areas to concentrate on to maintain the objectives of organic gardening. These includes: soil management which is dealt with by using organic fertilizer; weed management which is managed by manual labor and use of organic ground coverings; and lastly pest control which is dealt with by promoting beneficial insects and companion planting. These are the top key strategies for becoming an organic gardener. Proper knowledge is essential in organic gardening and requires simple fundamental lessons to get reliable results.

Organic Gardening magazine editor, Therese Ciesinski, describes organic gardening as:

"Organic gardening is more than simply avoiding synthetic pesticides and fertilizers. It is about observing nature's processes, and emulating them in your garden as best you can. And the most important way to do that is to understand the makeup of your soil and to give it what it needs. If anything could be called a 'rule' in organic gardening, it's this: feed the soil, not the plant."

Step by Step Process to Start an Organic Garden

The following is a step by step process on how to start an organic garden:

1. Prepare the Soil: The soil is the most important thing or resource when it comes to organic gardening. It is achieved through continuous addition of organic matter to the soil by using locally available resources in every possible aspect. If you want your plants to be healthy, you will need to thoroughly prepare the soil on which they will grow on.

Just like human beings, plants require food and the food in this case comes from the soil. Therefore, you need to ensure that your plants get plenty of fresh nutrients. Proper soil conditioning will give your plants all the nutrients they need.

Chemical soil treatments not only destroy the soil composition but they also harm the important microorganisms, worms, and bacteria in the soil. To begin, you will have to test the soil PH. You can do this by buying a home testing kit or simply collect some soil samples and send them to the local agricultural extension office for proper testing and analysis.

2. Make Good Compost: As you wait for soil sample results, you can make the compost. Compost helps in providing plants with nutrients, helps conserve water, helps in the reduction of weeds and helps in keeping food as well as yard waste out of landfills.

Compost can be obtained or made from locally available resources such as leaves, grass trimmings, yard garbage/remains, and kitchen waste. Alternatively, compost is readily available for purchase from mulch suppliers or organic garden centers. You can use these steps in making compost:

- Measure out space that is at least three square feet.
- Get a pile of natural dead plants or leaves.
- Add alternating layers of leaves, garden trimmings (carbon) and nitrogen (green) materials, for instance, kitchen leftovers and manure. Put a layer or separate them with a layer of soil.
- Cover the pile with about 4-6 inches of soil.
- Turn the pile every time a new layer is added to the mixture. During this process, ensure that water is added to keep the mixture moist to enhance microorganism activity.
- The compost should not smell but in case it does, add some more dry leaves, saw dust or straw and then turn it regularly. Do this for about four weeks or a month and you will have good compost needed for your organic garden.

3. Prepare your Garden: Once you are waiting for the compost to be ready, the next step is to prepare your garden. After receiving a go ahead from your local agricultural extension officer concerning the right soil type, it is now time to prepare the garden area. Using available gardening tools you can carefully prepare your garden. However, it is important to ensure that you do not completely destroy the soil.

4. Choose the Right Plants: Once you are through with preparing your garden, the next step is to select the right plants for your garden. Soil sampling and testing will come in handy at this stage.

It is important to choose plants that will thrive well in specific micro-conditions of your soil type. Carefully choose plants that will thrive well in different spots in your garden in terms of moisture, light, and drainage as well as the soil quality. Remember the healthier your plants are, the more and more resistant your crops will be to attackers.

Another mechanism for growing organically is to select plants suited to the garden. Crops well adapted to the gardens climate and conditions are better able to grow with minimal input. Also, growing crops well adapted to the site ensure greater natural defenses. Meaning, little attention and input is required for boosting crop productivity.

When purchasing seedlings, ensure that you go for plants that are raised without synthetic chemicals or pesticides. Your local farmers market is a good place to do your purchases. You will not only find a variety of plants but plant varieties that are best suited for your local area. Carefully select plants that look healthy and are without overcrowded roots.

5. Plant the Crops in Beds: When planting your crops ensure that you plant them in wide beds. Planting them in beds prevents you from walking on them and destroying the soil surface when harvesting or when cutting the flowers. Additionally, grouping crops helps in reducing weeding, wastage of water and makes it easier for you to apply compost. It also enables the plants to utilize the available nutrients and water. Ensure that there is adequate space between the rows. This helps in promoting air circulation which helps in repelling fungal attacks.

6. Water the Crops: Once you have planted your crops, the next step is to water them. It is good to water the plants immediately after planting them to give them the much-needed water to enable them to continue growing. You can also water them every morning. It is recommended to water your pants in the morning because there are no strong winds, mornings are cool and the water lost as a result of evaporation is immensely reduced. Experts recommend considerable, infrequent watering for plants that are already established.

7. Weeding: According to CSU, Weeds reduce crop yield by competing for water, light, soil nutrients, and space. In agricultural crops, weeds can reduce crop quality by contaminating the commodity. They can serve as hosts for diseases or provide shelter for insects to overwinter. It may be a hard work to pull out weeds by hand but then its a good exercise that helps you to get some fresh air.

8: Provide Nutrients to your Plants: When you do organic gardening, you need to look out for eco-friendly ways to protect your plants from toxic pesticides and fertilizers. So, you need to make sure that your plants get enough light, nutrients and moisture that help them to grow better. Also, a diverse garden would help to prevent pests, by limiting the amount of one type of plant and gives boost to biodiversity.

Instead of using chemical pesticides, organic gardening emphasizes on promoting beneficial insects and companion planting. Organic farmers need not to eliminate insects and diseases by using chemicals. Rather, pest control is done through keeping pests and diseases below damaging levels. One of the major mechanisms is by promoting beneficial insects and pest predators such as bats, birds, lizards, toads, and spiders.

The key to succeeding in this area is to grow a wide variety of companion crops that supports the ecological niche of these species. Avoiding the use of synthetic pesticides as well ensures their survival. Nipping off infected/infested leaves or buds, uprooting infected crops, crop rotation, and handpicking insect pests and eggs are excellent methods for controlling pest populations. Maintaining garden cleanliness is another effective tool for organic pest control.

Natural sprays or pesticides can equally be used in addition to the cultural pest control methods. They are readily available from organic garden centers and their products contain the bacterium Bacillus, neem oil, and minerals like copper. Their ability to break down quickly has promoted their wide usage in place of the synthetic chemical pesticides. Besides, there are some vegetable/fruit pests and diseases that are beyond natural and organic control, thereby requiring the use of natural sprays.

9. Use of Organic Fertilizers: As much as organic matter and compost will improve water and nutrient retention in the soil, the supply of all the required nutrients for healthy and productive growth is limited. Besides compost, organic gardening requires additional fertilizers drawn from natural sources such as: plant products like wood ash; natural deposits for example rock phosphate; and animal byproducts as well as manures.

Agricultural lime is another natural product frequently added to soil to improve its quality. It is produced from naturally occurring limestone and is added to the soil to optimize pH if the soil is too acidic. Soil pH levels vary from one locality to another. Local extension offices usually provide guidance on soil pH level testing within their jurisdiction areas. However, most soils do not need additional liming.

10. Taking Care of your Organic Garden: There are important maintenance tips for your organic garden if you want to maintain or harvest healthy plants. Some of these practices include:

- Mulching.

- Watering in the mornings.

- Use of compost.

- Use of natural manure especially from animals that do not eat meat.

- Weeding at the right time.

- Prune regularly to allow for proper aeration and effective use of nutrients including light.

Various Benefits of Organic Gardening

1. Reduces the Amount of Pesticides you and your Family Consume: Organic gardening focuses on the use of only natural products to grow plants. This means that there is no use of pesticides. Therefore, the crops obtained from this type of gardening are free of pesticides and other chemicals. You will live a healthy life without worrying about consuming chemicals through consuming crops.

Organic gardening promises good health because the produce is free from toxic ingredients and other synthetically enhanced chemicals. The fruits and vegetables grown in the organic garden do not have the chemical residues which enter the body when eaten. Organic veggies are also proven to contain higher mineral and vitamin contents compared to those grown using pesticides, herbicides, and chemical fertilizers.

Furthermore, complementary medicine professionals affirm that there are high nutrient concentrations such as Vitamins C and D in organic food products. Organic gardening also adds an extra benefit of exercising the body, especially from manual labor including planting, weeding, and harvesting. Thus, gardening can be an enjoyable and meaningful way to boost physical exercise. The outdoor environment also offers a refreshing means of connecting with nature, sunshine and fresh air, acting as a stress reliever.

2. Helps in Conserving the Environment: Use of chemicals on crops is one way in which we pollute the environment. The chemicals sprayed or applied on the crops seethe through the soil into the water. This puts the microorganisms at risk. When spraying the crops, the wind carries away the chemicals into the atmosphere, and this is air pollution. Therefore, embracing organic gardening is one of the best ways of maintaining a healthy environment.

Choosing organic gardening massively aids in environmental preservation. Growing vegetables and fruits in a natural way will not only ensure healthy produce but also promote a friendly and toxicity free environment. Organic gardening is the surest way of preserving a healthy and green environment by ensuring ecological balance and minimal disturbance to the natural environment.

It ensures birds, small animals, and beneficial insects are free from chemical harm. Organic matter used to prepare the soil also helps to improve the soil quality. Therefore, organic gardening offers the most beneficial outcome for good environmental health. Putting an end to chemical fertilizers and pesticides that leach into the ground and find way into water supply is only a possibility through organic gardening.

3. Reduces Greenhouse Gas Emissions: Crops that contain chemicals release these chemicals into the atmosphere in the form of greenhouse gasses. These gasses combined with other impurities in the air pose a risk to the air we breathe in. Additionally, these gasses also contribute immensely to global warming, and we all know the negative effects of global warming.

4. Better Taste: Organic vegetables and fruits contain scent and taste which are simply better due to their natural growth compared to those grown commercially. To a very large extent, veggies and fruits grown commercially cannot bear or beat the natural flavors of those that are grown organically. Fresh vegetables and fruits from the garden have always tasted better and have natural flavor.

5. Saves Money: A tremendous way of saving money is by growing your own organic vegetable garden. Saving money is something that every person wants to do but it can only be realized by

undertaking small initiatives like growing own veggies and fruits. Through organic gardening, one can save up to 50% of the money used to buy fruits and veggies at supermarkets as well as other perishable stores.

To conclude, organic gardening is beneficial to the people and the environment. However, learning how to start an organic is a very important step in ensuring that you conserve the environment and that you grow your own delicious fresh produce. To give your crops the needed nourishment, you can use natural fertilizers, for example, seaweed extracts, fish emulsion or manures obtained from animal droppings especially cow and chicken droppings that is readily available or bought from your local garden center.

Organic Food

Organic food refers to the fresh or processed food produced by organic farming methods. Organic food is grown without the use of synthetic chemicals, such as human-made pesticides and fertilizers, and does not contain genetically modified organisms (GMOs). Organic foods include fresh produce, meats, and dairy products as well as processed foods such as crackers, drinks, and frozen meals. The market for organic food has grown significantly since the late 20th century, becoming a multibillion dollar industry with distinct production, processing, distribution, and retail systems.

Policy

Although organic food production began as an alternative farming method outside the mainstream, it eventually became divided between two distinct paths: (1) small-scale farms that may not be formally certified organic and thus depend on informed consumers who seek out local, fresh, organically grown foods; and (2) large-scale certified organic food (fresh and processed) that is typically transported large distances and is distributed through typical grocery store chains. If consumers know their local farmer and trust the farmer's production methods, they may not demand a certification label. On the other hand, organic food produced far away and shipped is more likely to require a certification label to promote consumer trust and to prevent fraud, which exemplifies how national certification regulations are most beneficial.

Farmers' market Display of organic produce at an outdoor farmers' market.

A regulatory framework is most important when consumers and farmers are geographically separated, and such a framework is likely to cater to larger-scale producers who participate in a more industrial system. This regulatory approach does not necessarily match consumers' assumptions about organic food production, which typically include images of small family farms and the humane treatment of animals. In general, regulations surrounding organic food do not address more complex social concerns about family farms, farmworker wages, or farm size, and organic policy in some places does little to address animal welfare.

Organic food policies were created largely to provide a certification system with specific rules regarding production methods, and only products that follow the guidelines are allowed to use the certified organic labels. In the United States, the Organic Foods Production Act of 1990 began the process of establishing enforceable rules to mandate how agricultural products are grown, sold, and labeled. The regulations concerning organic food and organic products are based on a National List of Allowed and Prohibited Substances, which is a critical aspect of certified organic farming methods. The United States Department of Agriculture (USDA) regulates organic production through its National Organic Program (NOP), which serves to facilitate national and international marketing and sales of organically produced food and to assure consumers that USDA certified organic products meet uniform standards. To this end, NOP established three specific labels for consumers on organic food products: "100% organic," "organic," or "made with organic," which signify that a product's ingredients are 100 percent, at least 95 percent, or 70 percent organic, respectively. Noncertified products cannot use the USDA organic seal, and violators face significant fines and penalties.

Organic regulations vary by country, some of the most comprehensive rules being seen in Europe. Objectives of organic farming in the European Union (EU) include respecting nature's biological systems and establishing a sustainable management system; using water, soil, and air responsibly; and adhering to animal welfare standards that meet species-specific behavioral needs. In addition, principles of organic production in the EU are based on designing and managing farms to promote ecological systems and on using natural resources within the farming system. These policy goals go far beyond a defined listing of prohibited materials in organic production.

Environment

The overall impacts of organic agriculture are beneficial to the environment. Certified organic production methods prohibit the use of synthetic fertilizers and pesticides, thus reducing chemical runoff and the pollution of soils and watersheds. Smaller-scale organic farming often is associated with significant environmental benefits, owing to the use of on-farm inputs, such as fertilizers derived from compost created on-site. By comparison, large-scale organic farms often require inputs generated off-site and may not employ integrated farming methods. These operations may buy specific allowable inputs, such as fish emulsion or blood meal to use as fertilizer rather than working within the farm to increase soil fertility. While this decrease in synthetic chemical use benefits the environment compared with industrial agriculture, these methods may not promote long-term sustainability, since off-farm inputs usually require greater fossil fuel use than on-farm inputs.

Society

Social concerns related to organic food include higher costs to consumers and geographic variations in demand. Organic food usually is more expensive for consumers than conventionally produced food because of its more labour-intensive methods, the costs of certification, and the decreased reliance on chemicals to prop up crop yields. This often translates into unequal access to organic food. Research indicates that greater wealth and education levels are correlated with organic food purchases. Further, there are trends in some lower-income countries to produce certified organic crops solely for export to wealthier countries. This sometimes generates a situation in which the farmers themselves cannot afford to buy the organic foods they are producing. While this strategy may bring economic gain in the short term, it is a concern when farmers are forced out of producing food crops that feed their local communities, thus increasing food insecurity.

Certified organic agriculture has also become a big business in many places, with larger farming operations playing a key role in national and global certified organic food markets. Given economies of scale, big food-processing companies often buy from a single farming operation that produces organic crops on thousands of acres, rather than from many smaller farms that each grow on smaller acreages, a practice that effectively limits the participation of smaller farmers in these markets. There also is disparity among farmers, since the organic certification process can be prohibitively expensive to some smaller-scale farmers. Although certification subsidies exist in some places, such farmers often opt to sell directly to consumers at farmers' markets, for example, and may decide to forgo organic certification altogether.

Overall, organic food has grown in popularity, as consumers have increasingly sought and purchased foods that they perceive as being healthier and grown in ways that benefit the environment. Indeed, consumers typically buy organic food in order to reduce their exposure to pesticide residues and GMOs. Further, some research shows that organically produced crops have higher nutritional content than comparable nonorganic crops, and some people find organic foods to be tastier. The question remains, however, whether organic food shipped in from across the globe is truly a sustainable method of food production. Certainly organically produced food from a local farmer who employs an integrated whole-farm approach is fairly environmentally sustainable, though the economic sustainability of such an endeavour can be challenging. Although humans must decrease their reliance on fossil fuels to combat climate change, many organic policies do little to address the issue of sustainability, focusing instead on the strict list of prohibited substances, rather than a comprehensive long-term view of farming and food.

Chemical Composition

With respect to chemical differences in the composition of organically grown food compared with conventionally grown food, studies have examined differences in nutrients, antinutrients, and pesticide residues. These studies generally suffer from confounding variables, and are difficult to generalize due to differences in the tests that were done, the methods of testing, and because the vagaries of agriculture affect the chemical composition of food; these variables include variations in weather (season to season as well as place to place); crop treatments (fertilizer, pesticide, etc.); soil composition; the cultivar used, and in the case of meat and dairy products, the parallel variables in animal production. Treatment of the foodstuffs after initial gathering (whether milk is pasteurized or raw), the length of time between harvest and analysis, as well as conditions of transport and storage, also affect the chemical composition of a given item of food. Additionally, there is evidence that organic produce is drier than conventionally grown produce; a higher content in any chemical category may be explained by higher concentration rather than in absolute amounts.

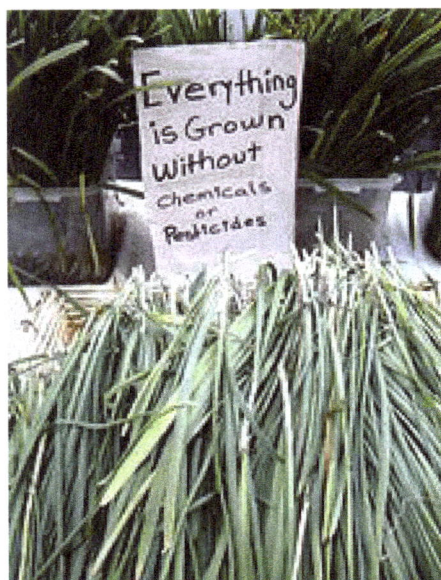

Organic vegetables at a farmers' market.

Nutrients

Many people believe that organic foods have higher content of nutrients and thus are healthier than conventionally produced foods. However, scientists have not been equally convinced that this is the case as the research conducted in the field has not shown consistent results.

A 2009 systematic review found that organically produced foodstuffs are not richer in vitamins and minerals than conventionally produced foodstuffs. The results of the systematic review only showed a lower nitrogen and higher phosphorus content in organic produced compared to conventionally grown foodstuffs. Content of vitamin C, calcium, potassium, total soluble solids, copper, iron, nitrates, manganese, and sodium did not differ between the two categories.

A 2012 survey of the scientific literature did not find significant differences in the vitamin content of organic and conventional plant or animal products, and found that results varied from study to study. Produce studies reported on ascorbic acid (vitamin C) (31 studies), beta-carotene (a precursor for vitamin A) (12 studies), and alpha-tocopherol (a form of vitamin E) (5 studies) content; milk studies reported on beta-carotene (4 studies) and alpha-tocopherol levels (4 studies). Few studies examined vitamin content in meats, but these found no difference in beta-carotene in beef, alpha-tocopherol in pork or beef, or vitamin A (retinol) in beef. The authors analyzed 11 other nutrients reported in studies of produce. A 2011 literature review found that organic foods had a higher micronutrient content overall than conventionally produced foods.

Similarly, organic chicken contained higher levels of omega-3 fatty acids than conventional chicken. The authors found no difference in the protein or fat content of organic and conventional raw milk.

A 2016 systematic review and meta-analysis found that organic meat had comparable or slightly lower levels of saturated fat and monounsaturated fat as conventional meat, but higher levels of both overall and n-3 polyunsaturated fatty acids. Another meta-analysis published the same year found no significant differences in levels of saturated and monounsaturated fat between organic and conventional milk, but significantly higher levels of overall and n-3 polyunsaturated fatty acids in organic milk than in conventional milk.

Anti-nutrients

The amount of nitrogen content in certain vegetables, especially green leafy vegetables and tubers, has been found to be lower when grown organically as compared to conventionally. When evaluating environmental toxins such as heavy metals, the USDA has noted that organically raised chicken may have lower arsenic levels. Early literature reviews found no significant evidence that levels of arsenic, cadmium or other heavy metals differed significantly between organic and conventional food products. However, a 2014 review found lower concentrations of cadmium, particularly in organically grown grains.

Phytochemicals

A 2014 meta-analysis of 343 studies on phytochemical composition found that organically grown crops had lower cadmium and pesticide residues, and 17% higher concentrations of polyphenols than conventionally grown crops. Concentrations of phenolic acids, flavanones, stilbenes, flavones, flavonols, and anthocyanins were elevated, with flavanones being 69% higher. Studies on phytochemical composition of organic crops have numerous deficiencies, including absence of standardized measurements and poor reporting on measures of variability, duplicate or selective reporting of data, publication bias, lack of rigor in studies comparing pesticide residue levels in organic and conventional crops, the geographical origin of samples, and inconsistency of farming and post-harvest methods.

Pesticide Residues

The amount of pesticides that remain in or on food is called pesticide residue. In the United States, before a pesticide can be used on a food crop, the U.S. Environmental Protection Agency must determine whether that pesticide can be used without posing a risk to human health.

A 2012 meta-analysis determined that detectable pesticide residues were found in 7% of organic produce samples and 38% of conventional produce samples. This result was statistically heterogeneous, potentially because of the variable level of detection used among these studies. Only three studies reported the prevalence of contamination exceeding maximum allowed limits; all were from the European Union. A 2014 meta-analysis found that conventionally grown produce was four times more likely to have pesticide residue than organically grown crops.

The American Cancer Society has stated that no evidence exists that the small amount of pesticide residue found on conventional foods will increase the risk of cancer, although it recommends thoroughly washing fruits and vegetables. They have also stated that there is no research to show that organic food reduces cancer risk compared to foods grown with conventional farming methods.

The Environmental Protection Agency maintains strict guidelines on the regulation of pesticides by setting a tolerance on the amount of pesticide residue allowed to be in or on any particular food.

Although some residue may remain at the time of harvest, residue tend to decline as the pesticide breaks down over time. In addition, as the commodities are washed and processed prior to sale, the residues often diminish further.

Bacterial Contamination

A 2012 meta-analysis determined that prevalence of *E. coli* contamination was not statistically significant (7% in organic produce and 6% in conventional produce). While bacterial contamination is common among both organic and conventional animal products, differences in the prevalence of bacterial contamination between organic and conventional animal products were also statistically insignificant.

Organic Meat Production Requirements

United States

Organic meat certification in the United States requires farm animals to be raised according to USDA organic regulations throughout their lives. These regulations require that livestock are fed certified organic food that contains no animal byproducts. Further, organic farm animals can receive no growth hormones or antibiotics, and they must be raised using techniques that protect native species and other natural resources. Irradiation and genetic engineering are not allowed with organic animal production. One of the major differences in organic animal husbandry protocol is the "pasture rule": minimum requirements for time on pasture do vary somewhat by species and between the certifying agencies, but the common theme is to require as much time on pasture as possible and reasonable.

Health and Safety

There is little scientific evidence of benefit or harm to human health from a diet high in organic food, and conducting any sort of rigorous experiment on the subject is very difficult. A 2012 meta-analysis noted that "there have been no long-term studies of health outcomes of populations consuming predominantly organic versus conventionally produced food controlling for socioeconomic factors; such studies would be expensive to conduct." A 2009 meta-analysis noted that "most of the included articles did not study direct human health outcomes. In ten of the included studies (83%), a primary outcome was the change in antioxidant activity. Antioxidant status and activity are useful biomarkers but do not directly equate to a health outcome. Of the remaining two articles, one recorded proxy-reported measures of atopic manifestations as its primary health outcome, whereas the other article examined the fatty acid composition of breast milk and implied possible health benefits for infants from the consumption of different amounts of conjugated linoleic acids from breast milk." In addition, as discussed above, difficulties in accurately and meaningfully measuring chemical differences between organic and conventional food make it difficult to extrapolate health recommendations based solely on chemical analysis.

As of 2012, the scientific consensus is that while "consumers may choose to buy organic fruit, vegetables and meat because they believe them to be more nutritious than other food the balance of current scientific evidence does not support this view." The evidence of beneficial health effects of organic food consumption is scarce, which has led researchers to call for more long-term studies.

In addition, studies that suggest that organic foods may be healthier than conventional foods face significant methodological challenges, such as the correlation between organic food consumption and factors known to promote a healthy lifestyle. When the American Academy of Pediatrics reviewed the literature on organic foods in 2012, they found that "current evidence does not support any meaningful nutritional benefits or deficits from eating organic compared with conventionally grown foods, and there are no well-powered human studies that directly demonstrate health benefits or disease protection as a result of consuming an organic diet."

Consumer Safety

Pesticide Exposure

The main difference between organic and conventional food products are the chemicals involved during production and processing. The residues of those chemicals in food products have dubious effects on human health. All food products on the market including those that contain residues of pesticides, antibiotics, growth hormones and other types of chemicals that are used during production and processing are said to be safe.

Claims of improved safety of organic food has largely focused on pesticide residues. These concerns are driven by the facts that "(1) acute, massive exposure to pesticides can cause significant adverse health effects; (2) food products have occasionally been contaminated with pesticides, which can result in acute toxicity; and (3) most, if not all, commercially purchased food contains trace amounts of agricultural pesticides." However, as is frequently noted in the scientific literature: "What does not follow from this, however, is that chronic exposure to the trace amounts of pesticides found in food results in demonstrable toxicity. This possibility is practically impossible to study and quantify;" therefore firm conclusions about the relative safety of organic foods have been hampered by the difficulty in proper study design and relatively small number of studies directly comparing organic food to conventional food.

Additionally, the Carcinogenic Potency Project, which is a part of the US EPA's Distributed Structure-Searchable Toxicity (DSSTox) Database Network, has been systemically testing the carcinogenicity of chemicals, both natural and synthetic, and building a publicly available database of the results for the past ~30 years. Their work attempts to fill in the gaps in our scientific knowledge of the carcinogenicity of all chemicals, both natural and synthetic, as the scientists conducting the Project described in the journal, *Science*, in 1992.

Toxicological examination of synthetic chemicals, without similar examination of chemicals that occur naturally, has resulted in an imbalance in both the data on and the perception of chemical carcinogens. Three points that we have discussed indicate that comparisons should be made with natural as well as synthetic chemicals:

1. The vast proportion of chemicals that humans are exposed to occur naturally. Nevertheless, the public tends to view chemicals as only synthetic and to think of synthetic chemicals as toxic despite the fact that every natural chemical is also toxic at some dose. The daily average exposure of Americans to burnt material in the diet is ~2000 mg, and exposure to natural pesticides (the chemicals that plants produce to defend themselves) is ~1500 mg. In comparison, the total daily exposure to all synthetic pesticide residues combined is ~0.09 mg. Thus, we estimate that 99.99%

of the pesticides humans ingest are natural. Despite this enormously greater exposure to natural chemicals, 79% (378 out of 479) of the chemicals tested for carcinogenicity in both rats and mice are synthetic (that is, do not occur naturally).

2. It has often been wrongly assumed that humans have evolved defenses against the natural chemicals in our diet but not against the synthetic chemicals. However, defenses that animals have evolved are mostly general rather than specific for particular chemicals; moreover, defenses are generally inducible and therefore protect well from low doses of both synthetic and natural chemicals.

3. Because the toxicology of natural and synthetic chemicals is similar, one expects (and finds) a similar positivity rate for carcinogenicity among synthetic and natural chemicals. The positivity rate among chemicals tested in rats and mice is ~50%. Therefore, because humans are exposed to so many more natural than synthetic chemicals (by weight and by number), humans are exposed to an enormous background of rodent carcinogens, as defined by high-dose tests on rodents. We have shown that even though only a tiny proportion of natural pesticides in plant foods have been tested, the 29 that are rodent carcinogens among the 57 tested, occur in more than 50 common plant foods. It is probable that almost every fruit and vegetable in the supermarket contains natural pesticides that are rodent carcinogens.

While studies have shown via chemical analysis, as discussed above, that organically grown fruits and vegetables have significantly lower pesticide residue levels, the significance of this finding on actual health risk reduction is debatable as both conventional foods and organic foods generally have pesticide levels well below government established guidelines for what is considered safe. This view has been echoed by the U.S. Department of Agriculture and the UK Food Standards Agency.

A study published by the National Research Council in 1993 determined that for infants and children, the major source of exposure to pesticides is through diet. A study published in 2006 by Lu et al. measured the levels of organophosphorus pesticide exposure in 23 school children before and after replacing their diet with organic food. In this study it was found that levels of organophosphorus pesticide exposure dropped from negligible levels to undetectable levels when the children switched to an organic diet, the authors presented this reduction as a significant reduction in risk. The conclusions presented in Lu et al. were criticized in the literature as a case of bad scientific communication.

More specifically, claims related to pesticide residue of increased risk of infertility or lower sperm counts have not been supported by the evidence in the medical literature. Likewise the American Cancer Society (ACS) has stated their official position that "whether organic foods carry a lower risk of cancer because they are less likely to be contaminated by compounds that might cause cancer is largely unknown." The risks from microbiological sources or natural toxins are likely to be much more significant than short term or chronic risks from pesticide residues.

Microbiological Contamination

Organic farming has a preference for using manure as fertilizer, compared to conventional farming in general. This practise seems to imply an increased risk of microbiological contamination, such as *E. coli* O157:H7, from organic food consumption, but reviews have found little evidence that actual incidence of outbreaks can be positively linked to organic food production. The 2011 Germany E. coli O104:H4 outbreak, however, was blamed on organic farming of bean sprouts.

Environmental safety

From an environmental perspective, fertilizing, overproduction and the use of pesticides in conventional farming has caused, and is causing, enormous damage worldwide to local ecosystems, biodiversity, groundwater and drinking water supplies, and sometimes farmer's health and fertility. Outcomes from organic farming, however, are uncertain for their environmental benefits, and have limits for transforming the food system, where reducing food waste and dietary changes might provide greater benefits.

The Benefits of Organic Food

How your food is grown or raised can have a major impact on your mental and emotional health as well as the environment. Organic foods often have more beneficial nutrients, such as antioxidants, than their conventionally-grown counterparts and people with allergies to foods, chemicals, or preservatives often find their symptoms lessen or go away when they eat only organic foods.

Organic produce contains fewer pesticides. Chemicals such as fungicides, herbicides, and insecticides are widely used in conventional agriculture and residues remain on (and in) the food we eat. Organic food is often fresher because it doesn't contain preservatives that make it last longer. Organic produce is often (but not always, so watch where it is from) produced on smaller farms near where it is sold.

Organic farming is better for the environment. Organic farming practices reduce pollution, conserve water, reduce soil erosion, increase soil fertility, and use less energy. Farming without pesticides is also better for nearby birds and animals as well as people who live close to farms. Organically raised animals are NOT given antibiotics, growth hormones, or fed animal byproducts. Feeding livestock animal byproducts increases the risk of mad cow disease (BSE) and the use of antibiotics can create antibiotic-resistant strains of bacteria. Organically-raised animals are given more space to move around and access to the outdoors, which help to keep them healthy.

Organic meat and milk are richer in certain nutrients. Results of a 2016 European study show that levels of certain nutrients, including omega-3 fatty acids, were up to 50 percent higher in organic meat and milk than in conventionally raised versions. Organic food is GMO-free. Genetically Modified Organisms (GMOs) or genetically engineered (GE) foods are plants whose DNA has been altered in ways that cannot occur in nature or in traditional crossbreeding, most commonly in order to be resistant to pesticides or produce an insecticide.

Organic Vegetable Production

There are many reasons to consider growing organic vegetables. Organic production is a system that lends itself well to small-scale and part-time farming operations. Additionally, although the cost of certification and the time and labor involved in managing the system are high, returns have the potential to be high where markets are well developed for organic products.

Producing organic vegetables will require detailed recordkeeping and working very closely with a certifying agency. You may use the certifying agency's record keeping example or develop your own, if it is accepted by your agency. Before using any product, check with your certifying agency to make sure the product is suitable for organic production (especially if it is not on the certifying agency's list).

Consumer demand for organic produce is high and production has expanded significantly in recent years. There are almost 13,000 certified organic farms in the United States with a sales value of over $3.1 billion. In the northeastern United States, more than 3,200 farms are certified organic and have a sales value of over $300 million. If you are considering organic vegetable production, however, you should carefully evaluate the market regional demand for organic products and then decide which marketing channels will best meet the needs of your consumers. Marketing organic vegetables involves significant transportation and labor costs because most of your potential consumers are likely to be located in higher-income urban and suburban areas, which may be a considerable distance from your farm.

The U.S. Department of Agriculture (USDA) regulates the use of the term "organic." In order to become certified organic, a grower uses production and handling practices in accordance with the National Organic Program (NOP) and becomes certified by a USDA-accredited certifying agency. Growers whose gross income from organic production is $5,000 or less are exempt from this rule. In this case, production and handling practices in accordance with the NOP are still used and other restrictions regarding labeling and combination with other organic products apply. It takes a minimum of three years to transition from nonorganic production to certified organic production. During this period products cannot be labeled as organic and may not command the higher prices usually associated with organic products. Growing organic requires detailed recordkeeping and more management and planning time than other production systems.

The following summarizes the portions of the organic regulations that apply to vegetable production.

Organic Certification Process

Organic certification is a process where a third party accredited by the USDA ensures or certifies that vegetables were grown following the production and handling practices required by the NOP.

Steps to Receiving Organic Certification

1. Select an accredited certifying agency and request an application packet.

2. Develop an Organic System Plan (OSP) for areas where organic vegetables will be grown and implement the plan each year during the three-year transition period.

3. Submit the application packet to the certification agency in the third year of the transition.

4. Schedule an inspection of the field(s) where certification is requested.

5. Wait for a review of the inspection records by the certification staff.

6. Receive notification of organic certification or information on why certification was denied.

Certification lasts for one year, after which time you must become recertified. The amount of paper work required generally decreases after the first year because the OSP is already developed and just needs to be updated.

Accredited Certifying Agencies

Any certifying agency accredited with the USDA is authorized to award organic certification to any grower. As of 2014, 82 certifying agencies were accredited worldwide: 49 in the United States and 33 abroad.

Considerations for selecting a certifying agency can include the fees associated with certification, accreditation to standards for export, if needed, and additional services. Costs for becoming certified vary depending on the certifying agency and generally include certification fees, the cost of inspection, administrative fees, and additional assessments. When producing an organic crop for export, it is important to verify that the accreditation extends to the importing country. Additional services can include educational events and materials and membership benefits.

Organic System Plan

An Organic System Plan (OSP) is a thorough record describing the practices and procedures used on the farm to achieve and maintain the requirements of the NOP. An OSP is unique to a particular operation. Many certifying agencies have templates for developing an OSP, or you can develop your own.

An OSP Contains Six Elements

1. A description of practices and procedures, including frequency of use.

2. A list of each substance, including composition and commercial availability, to be used in the operation and where it will be used.

3. A description of monitoring techniques, including frequency of use.

4. A description of the recordkeeping system. The records must be maintained for a minimum of five years and made available during business hours for inspection.

5. A description of the establishment and management of physical barriers or buffer zones on operations with both organic and nonorganic components. In addition, methods to avoid commingling of organic and nonorganic products and prevent contact with prohibited substances must be described.

6. Other information deemed necessary by the certifying agency to determine compliance with the NOP.

It is important to discuss practices and procedures (especially inputs) you plan on using with your certifying agency to verify that they are certifiable.

Production Practices

Plant Selection

The use of genetically modified organisms (GMOs) is prohibited in certified organic production. Seed, transplants, and other planting stock must be organically produced. Exceptions can be made when an "equivalent variety" of organic seed or planting stock is not commercially available in an organic form for a particular crop. Before these options or exceptions apply, you must have evidence that a minimum of three seed or planting stock sources were checked for organic forms.

The first exception allows you to use untreated, non-organic seed and planting stock, except for the production of edible sprouts. If untreated, conventionally produced seed or planting stock is not commercially available, seeds or planting stock treated with substances allowed according to the National List can be used. The National List catalogs allowable and prohibited substances in certified organic production. Seed, transplants, and planting stock that have been treated with prohibited substances can be used to produce an organic crop when the application of the substance is a requirement of federal or state phytosanitary regulations. A temporary variance can also be obtained for use of non-organic transplants when an unavoidable event, such as a fire, flood, or frost, has occurred. In addition, conventional planting stock that is used to produce perennial crops can be sold as organic after it has been managed using certified organic practices for a minimum of one year. Work closely with your certifying agency to ensure that exceptions and variances can be made without compromising your organic certification.

In addition to following the NOP criteria, select cultivars with good market characteristics. Also, consider cultivars with resistance or tolerance to insect and disease pests common in your area or field. When using transplants and other planting stock, buy certified disease-free stock when possible and only purchase from reputable suppliers.

Treated Lumber

Organic growers cannot use lumber treated with arsenate or other prohibited substances for new installations or as replacement for lumber in contact with soil or livestock.

Soil fertility

The goal of soil fertility management is to maintain or improve the condition of the soil and minimize soil erosion. This is done by using sound crop rotations, green manures and cover crops, plant and animal matter, and fertilizers or soil amendments allowable according to the National List. Soil testing should be used to determine pH and levels of phosphorus, potassium, calcium, and magnesium. The nutrient levels in the soil will indicate the amount of additional nutrients needed for optimal vegetable crop growth and development.

Plant and Animal Materials

Composted or uncomposted plant and animal materials not treated with prohibited substances can be applied to the soil. Composted plant materials can be incorporated into the soil as needed. Composted plant and animal materials can be incorporated into the soil as needed, provided the compost meets carbon-to-nitrogen (C:N) ratio and temperature requirements. Compost used must have an initial C:N ratio between 25:1 and 40:1. In addition, when using an in-vessel or static aerated pile composting system, the pile must reach temperatures between 131 and 170 °F for a minimum of 3 days. If using a windrow composting system, the pile temperature must be maintained between 131 and 170 °F for a minimum of 15 days and turned a minimum of five times during that time. Temperatures and turning must be documented. Composted materials can be tested to determine their nutrient content.

Uncomposted plant materials can also be used as needed in certified organic production.

Uncomposted animal Manure can be used only:

- On fields with crops not to be consumed by humans.

- If it is incorporated into the soil a minimum of 90 days before harvest, provided that the edible portion of the crop does not contact the soil.

- If it is incorporated into the soil a minimum of 120 days before harvest for a product that does come into contact with the soil.

Be aware that the regulations for using raw manure will likely be affected by the Food Safety Modernization Act. Using municipal sewage sludge is prohibited in certified organic production.

Fertilizers and Soil Amendments

Fertilizers and soil amendments allowable according to the National List are available to complement other fertility practices. In addition, mined materials of low solubility can be used to supply plant nutrients. Plant or animal ashes can also be used to improve soil fertility as long as they have not been combined or treated with a prohibited substance and are not themselves a prohibited substance. Be aware that some fertilizers and soil amendments labeled as "natural" or "organic" may not be allowed in organic production. Check with your certifying agency before applying any material to your fields.

One of the limitations to using organic fertilizers is that allowable fertilizers are sometimes difficult to find commercially, although this is improving as the industry grows. In addition, allowable fertilizers generally cost considerably more than synthetic fertilizers. They tend to be low in the amount of nutrients they supply and therefore may need to be applied in large amounts that can be difficult to manage. Lastly, organic fertilizers can be difficult to blend. It is best to use them to complement other fertility practices such as compost, cover crops, and animal manures.

Pest Management

Pests must be managed primarily using various management tactics to avoid them rather than pesticides to kill them. Preventive pest management options include use of cultural techniques, physical barriers, and biological controls. It is helpful to determine common pests of particular vegetable crops before planting. Cultural techniques, physical barriers, and/or biological controls can then be selected that manage potential pests.

Cultural techniques include good site and cultivar selection, proper moisture and nutrient management, sanitation, rouging, vector management, manipulating harvest schedules, crop rotation, using cover crops and green manures, mechanical cultivation, hand weeding, using trap crops, encouraging beneficial insects, mulching, and livestock grazing. Physical barriers include plastic or organic mulches and row covers. When using plastic mulches, they must be removed at the end of the growing or harvest season. Burning of crop resides is prohibited except when used for disease management or to promote seed germination.

If these strategies fail, allowable pesticides can be used.

Organic Fruit Production

Organic fruit production essentially excludes the use of many inputs associated with modern farming, most notably synthetic pesticides and fertilizers. To the maximum extent possible, organic farming systems rely upon crop rotations, crop residues, animal manures, legumes, green manures, off-farm organic wastes, mechanical cultivation, mineral- bearing rock powders and biological pest control. These components maintain soil productivity and tilth, supply plant nutrients and help to control insects, weeds and other pests.

Cultural Practices

Site Selection

All factors regarding site suitability for conventional fruit plantings (air and water drainage, etc.) apply even more and so to organic operations. While conventional farmers may fall back on chemical fertilizers and pesticides to compensate for poor site decisions, organic farmers largely give up such luxuries. The presence of certain weeds and forage species are also of particular concern to the organic farmer. Bermudagrass, Johnsongrass and several other species can be quite problematic to farmers and are difficult to control through nonchemical means.

Growing fruit crops offers an advantage to farmers interested in sustainable agriculture. Because fruit plantings are perennial, the soil may not require additional tillage or cultivation beyond that needed at establishment, thereby minimizing soil erosion. Because the potential for erosion is low, hillsides and other sites unsuitable for tillage agriculture can safely and successfully produce fruit crops.

Crop Selection

Environmental constraints (climate, presence of certain pests and diseases, suitable soils, etc.) can greatly impact the suitability of a given site or even a bioregion for organic production of a given fruit crop. Generally speaking, some perennial fruits are easier to grow organically than others. The small fruits (berries) for example, seem easier to grow organically than the tree fruits in almost all locations. Lastly, successful organic fruit growing may depend largely on whether the venture is for home production or for commercial sales.

Site Preparation

Attention to the details of site preparation can go a long way towards reducing weed and disease problems and assuring a vital planting through soil improvement.

In general, fruit crops do not require highly fertile soils for good production. In fact, highly fertile soils, rich in nitrogen, can promote too much vegetative growth at the expense of fruiting. A nutritionally balanced soil, proper oil pH and plentiful amounts of organic matter are the fundamentals of an organic fertility management plan for fruits. Preplant soil improvement for organic fruit plantings is typically accomplished through some combination of cover cropping/green manuring and the use of imported materials, which may include manures, compost, rock powders and organic wastes.

Of particular importance at this stage is the application of required amounts of lime or sulphur. The need for lime or sulphur is dependent on the crop to be grown and the results of soil testing. Adjusting the oil pH with lime (to raise the pH) or sulphur (to lower pH) is much more easily done before planting. Most fruit plants perform best around pH 6.5, although they tolerate a pH range between 5.5 and 7.2. Blueberries are an exception. They require an acid oil-ideally pH 4.8 to 5.2 soil testing should also be used to guide applications of manures and other rock powders to avoid nutrient imbalances.

Cover crops and green manures not only contribute to soil fertility but also can become part of an active plan for pre-plant weed suppression as smother crops. The basic strategy begins with ploughing under sod or other existing vegetation, planting a cover crop to suppress weed growth, tilling under the cover crop and setting plants. Sometimes several rotations of cover crops are used before setting plants.

Selection of specific cover crops and their management varies with location, depending on such factors as seasonal rainfall, soil type, soil erosion potential, available equipment and seed cost. For example, crotalaria probably performed less effectively than sesbania at different centre because it is poorly adapted to low, wet soils. Some other warm season cover crops that might be considered include aggressive maize varieties and forage soybean varieties.

Crops such as berries (strawberries) are often grown on raised beds to encourage drainage and reduce root rot disease problems.

Soil Solarization

Another technique for site preparation is soil solarization. The process involves placing transparent polyethylene plastic on moist soil during the hot summer months to increase soil temperatures to levels lethal to many pests. Solarization suppresses weeds and eliminates many potential soil disease and nematode problems. Soil solarization is somewhat expensive and is not frequently used to prepare sites for perennial fruits, with the possible exception of strawberries.

Orchard Floor Management and Mulching

The floor or inter row areas of perennial fruit plantings may be managed in a variety of ways, ranging from clean cultivation to various combinations of cover crops and organic mulches. In terms of controlling erosion at least, a system that results in full or close-to-full ground coverage is best. If the decision is made to plant a between row ground cover, it should consist of a species adapted to the region and to the fruit farmer's management plan. Where adapted, orchard grass and other cool season grasses are often recommended because they go dormant during the heat of the summer, thereby minimizing competition with the fruit crop for water. With proper fertility management these grasses may also provide plentiful material for mulch. Many warm season legumes are deep rooted and compete with the trees for water, they should not be allowed to grow under the tree canopy. However, legumes and the mulch made from leguminous ground covers can provide significant nitrogen to fruit trees or vines. Furthermore, even though they complete with the fruit plants for water, legumes (like the grasses) increase the water permeability of the soil. Increasing the organic matter content of the soil also increases the soil's moisture retentiveness.

Subterranean clover reseeds itself in early summer and dies back during hottest part of the growing season leaving a relatively thick, weed suppressive duff or mulch. This system has shown applicability in apple and peach orchards and for a variety of orchard crops. Subterranean clover is not adapted to climates where winter temperatures regularly drop below 0 °F.

Many researchers have also investigated the relationship between orchard floor management and pest control. In some place apple orchards peach orchards, a diverse mix of cover crop spices provided habitat and food for an array of beneficial organisms, resulting in a decrease of orchard pests.

Mustards, buckwheat, dwarf sorghum and various members of the Umbelliferae and Compositae families support substantial numbers of beneficial insects without attracting as many pests.

After a planting is established, mulching with organic materials such as straw, leaves or sawdust can provide significant weed suppression. If applied thickly enough or supplemented with sheets of paper or cardboard as the bottom layer, complete suppression might be achieved. Application of mulch varies somewhat by crop. In strawberries for instance, mulch might cover entire aisles between rows or only be placed next to the bed to inhibit encroachment by creeping weeds. In vineyards and orchards, mulch might be placed only under individual trees or vines or along the entire row, ideally extending to the drip line.

Generally, mulch should be kept well away from the trunk to reduce damage from voles. This is especially important in winter. Keeping the mulch 8-12 inches away from the trunk also reduces the likelihood of crown rot and other diseases in susceptible species most notably applies on certain rootstocks. In mulberries, sawdust mulch is commonly spread along the entire row with extra sawdust mounded around the canes, often to a depth of 8 or more inches. In addition to controlling weeds, mulching with organic materials improves the soil by enhancing soil aggregation and water availability.

Varietal Selection

Because the plants are perennial and represent a considerable investment in both time and money, it is important to start the fruit planting with the optimum varieties for location and markets.

Good information on varietal selection is available from Cooperative Extension, nurseries and local commercial farmers. It is also important to obtain clean planting stock. Buying from reputable nurseries that provide stock certified by certifying inspectors to be free of diseases and insect pests is recommended.

Genetic resistance refers to inheritable traits in the plants that inhibit disease and pest damage. Choosing genetically resistant cultivars is a very important control measure for organic farmers especially with regard to disease management. In some cases, such as bacterial spot in peaches, cultivar resistance is the best or only control.

Aphid resistant berries, *Phylloxera* resistant grape rootstocks, wooly aphid resistant apple rootstocks, mite resistant strawberry cultivars, and nematode resistant peach rootstocks are available. As important as this resistance is there is no cultivar of any fruit species with multiple insect pest resistance; therefore, means other than resistance will most likely have to be employed to protect fruit plant from a complex of several pest species.

Organic Fertilization Practices

Harvested fruit, being largely water, removes relatively few nutrients from the soil, compared to other crops. Therefore, a significant amount of the fertility needs of fruit crops can be met through cover crop management and organic mulches in systems which use them and by the application of lime and other slow release rock powders at the preplant stage. Still, supplementary fertilization is often required for optimum growth and production. Some information useful to planning a supplementary fertilization program for perennial fruits include:

- Organic farmers generally employ relatively non-soluble fertilizer materials, such as compost, manures, plant derived byproducts like cottonseed meal or animal byproducts like feather or blood meals. To insure adequate decomposition and timely nutrient release of these slowly available materials, early spring application is encouraged. Early application also reduces the tendency towards late season growth, which may result in winter damage.

- Surface application of organic fertilizer materials without incorporation is sometimes wasteful of the nitrogen contained in those materials. This is especially true of manures. However, incorporation by tillage could damage the roots of the fruit plants and increase erosion.

- To bypass some of the "problems" associated with slowly available organic materials, some organic fruit farmers choose soluble organic fertilizers such as fish emulsion, soluble fish powder or water soluble blood meal in some cases applying these as a foliar spray. Since these are relatively expensive, the prudent farmer seldom relies exclusively on such materials.

- Most organic fertilization programs focus on supplementing nitrogen as the key element since it is needed in the greatest amount for the crop. One way to determine a proper application rate for an organic fertilizer is to obtain a conventional chemical recommendation for the planting and calculate the approximate amount of an organic fertilizer required to meet it. For instance, if the recommendation for a berry crop is 160 kg of actual nitrogen per acre, a farmer choosing to use cottonseed meal (approximately 7% N) would need to apply about 2000 kg of that material to each acre. Such calculations are simplistic however and may lead to spending more on fertilizer inputs than in necessary. One reason is because biologically healthy soils fix and release greater amounts of nitrogen naturally than those which are not.

- Further more, when making fertilizer calculations based on nitrogen, the farmer needs to credit the estimated contributions made by legume cover crops and or leguminous mulches, where these are used. A cover crop of subterranean clover, properly fertilized and inoculated, can fix from 200 to 1000 kg of nitrogen per acre annually in a living mulch system.

- Basing application rates solely on nitrogen content can cause problems when the fertilizers themselves are imbalanced. Repeated use of poultry manure for example which is very high in phosphate, can lead to pollution problems and a zinc deficiency in the crop. These problems can be averted by regular soil testing and adjusting fertilizer selection and rates accordingly.

- The most reliable means for determining whether fertilization is adequate is to combine field observations with soil and tissue testing. Poor yields, unusual colouration of leaves, and poor plant growth are all clues to a possible nutritional imbalance or deficiency. For example, less than 10 inches annual elongation of the branches of most fruit trees probably indicates a nitrogen deficiency. A corky bark on certain apple varieties may indicate an over availability of manganese in the soil.

- A foliar analysis measures the nutrient content of the leaves and can identify a deficiency or excess well in advance of visible symptoms. It is more helpful than a soil test because the foliar analysis is a measure of what the plant is actually taking up, while soil analysis only measures what is in the soil, which may or may not be available to the plant. Annual foliar analyses generally provide the best guide for adjusting supplementary nitrogen fertilization.

Organic Weed Management

Research indicates that without some form of weed control in the fruit planting, crop yields and plant vigour will be greatly reduced. In organic farming, weed control is only one goal of a weed management system for perennial fruit crops. A good organic weed management plan should present a minimum erosion risk, provide a "platform" for the movement of farm equipment, not impact adversely on pest management or soil fertility, while minimizing weed competition for water and nutrients. This philosophy has already been demonstrated in discussions regarding three effective weed control tools: cover crops, mulches and soil solarization.

Organic Insect and Mite Pest Management

A major distinction between pest management in perennial fruit crops and in annual crops is that crop rotation is not an option (strawberries and to a lesser extent, brambles, are possible exceptions). The long term nature of fruit growing allows for the possible build up of a pest population over time. Conversely, it is also possible for such stable agricultural environments to sustain populations of beneficial organisms.

Plant Health and Vigour

Though it is sometimes overstated, maintaining the plant in general health and good vigour is important in pest management. For fruit plants, this adage is more applicable to indirect pests (those pests that feed on foliage, stems etc.) than to direct pests (pests that feed on the fruit). For instance, an apparently healthy plum tree may set a good crop of fruit, yet lose it all to the plum. On the other hand, the same tree might suffer significant defoliation by caterpillars early in the season; yet, if it is in good vigour, it can compensate and bounce back quickly still producing a marketable crop that year. There are some cases where general plant health and freedom from stress does impart a form of "resistance" not technically genetic resistance to certain pests. Two examples are apple trees in good vigour actually smothering with sap or casting out invading flathead apple tree borers and plants not suffering drought stress being much less attractive to grasshoppers.

Biological Control

Biological control refers to the use of living organisms to control the population of a pest. Examples of beneficial arthropods that have been used to control pests in fruit crops include the predatory mites *Phytoseiulus persimilis* and *Metaseiulus occidentalis,* which attack spider mites; lady bird beetles and green lacewings which feed on aphids and *Trichogramma* wasps, which parasitize the eggs of several pests including codling moth. Many beneficial insects can be purchased from commercial insectaries and released in fruit plantings. More economical, however is the management of cover crops and adjacent vegetation as insect refugia to attract and sustain native populations of beneficials. As a rule, it appears that beneficial arthropods are not a complete control measure for direct fruit pests at least for commercial farmers who have a low damage threshold for fresh fruit. Usually, additional control measures are required.

Organic and Biorational Pesticides

Pesticides approved for certified organic production are usually derived from natural sources, break down rapidly in the field and appear to have minimal impact on the environment. Examples include botanical extracts from plants, insect growth regulators, synthetic pheromone treatments that cause mating disruption, soaps, oils, minerals such as sulphur dust and biological pesticides. The term "biorational pesticide" is used to describe pesticides organically acceptable or synthetic, which have minimal impact on beneficial insects and the environment.

Biorational pesticides are considered those providing the least toxic control that can be applied against a pest. Biological pesticides (biopesticides) are both organically acceptable and biorational. Biopesticides differ from biological control in part because they are formulated, labeled and applied like standard pesticides but also because the organisms involved do not reproduce significantly in the field. Biopesticides are highly specific and do not harm humans or beneficial insects. Several biopesticides can be used in fruit pest management. To be effective, they must be used at a specific time in the pest's development cycle.

The bacterium *Bacillus thuringiensis* (Bt) is an example of a commonly used biological insecticide. Bt is not as effective against lepidopterous pests that spend their larval stage feeding inside stems, crowns, trunks, or fruit, etc. (e.g., peach tree borer, codling moth, grape root borer, etc.). Other microbial insecticides include *Bacillus popilliae* for Japanese beetle grubs, a granulosis virus for codling moth, and insect parasitic nematodes for grubs and wireworms.

Botanical insecticides are formulated by extracting toxic compounds from plants that have pesticidal properties. They are naturally occurring, short-lived in the environment and do not leave harmful residues. However, many botanicals are broad spectrum poisons, affecting pests and beneficial organisms alike and are not always the biorational choice. Organic farmers, who are prohibited from using synthetic pesticides, frequently use botanical extracts. Some commonly used plant-derived insecticides are pyrethrum, rotenone, ryania and neem.

Specially formulated soaps that are high in fatty acids are effective against several soft bodied insects including aphids, whiteflies, leafhoppers and spider mites. Insecticidal soap penetrates the insect's body and disrupts the normal function of cells and their membranes, causing the contents to leak out.

Applying a thin layer of dormant oil to certain woody plants such as fruit trees, grapevines and bushes suppresses pests like leaf rollers, aphids, mites and scales by suffocating over wintering adults and eggs. Dormant oils should be applied prior to bud break and should never be mixed with sulphur because foliage damage may occur. So-called summer oils have a very low viscosity and can be used during the growing season to control aphids and mites without plant damage if label cautions are observed. Insect pheromones are chemicals produced by insects to help them communicate such things as mate availability and sexual receptivity. They are usually specific to a given insect species or genus.

Scientists have learned how to synthesize many of these pheromones and they are widely used for monitoring the emergence or simple presence of crop pests. This information is commonly used to time pesticide applications. New technology also allows pheromones to be used for mating disruption of certain pests. Mating disruption pheromones are available for the oriental fruit moth, codling moth and peach tree borer and grape berry moth.

Organic Disease Management

General Disorders and Cultural Controls

There are however, some types of disease problems, which are common to almost all temperate zone perennial fruit crops. For instance, because of the relatively soft nature and high sugar content of most mature or nearly mature fruits, fruit rots are common afflictions. The organic farmer can help to minimize the chances of fruit rots by allowing good air circulation and sunlight penetration into the interior plant canopy. Sunlight and circulating air help to dry leaf and fruit surfaces, thereby limiting fungal and bacterial infections. In tree crops, this would mean proper pruning and training techniques. In brambles and berries, reducing plant density helps. In grapes, discouraging rank vine growth and removing leaves that shade fruit clusters is beneficial. For all fruit crops, a site that allows for good air circulation should be chosen.

Another problem common to many fruit crops is root rots and intolerance of poorly drained soils. Most pear rootstocks and some apple rootstocks are relatively tolerant of heavy of poorly drained soils, but even these crops will succumb to persistently waterlogged conditions. *Prunus* species (peaches, plums, cherries, etc) are very intolerant of poorly drained soils and are generally susceptible to root rotting organisms common in such soils. Other cultural aids in minimizing disease can include such things as maintaining plants in good health and vigour, removal through prunings from the planting site, rouging out of diseased plants, removal of alternate hosts or inoculum sources for the disease organisms.

Organic Fungicides and Bactericides

Copper and sulphur compounds are the principal fungicides and bactericides used by organic farmers, but they have drawbacks. These materials can cause damage to plants if applied incorrectly. Sulphur is also lethal to some beneficial insects, spiders and mites and can set the stage for further pest problems. Long term use of copper fungicides can also lead to toxic levels of copper in the soil. Furthermore, these fungicides are typically inferior to synthetic alternatives, and have to be used on a protective schedule requiring frequent applications. Research on biofungicides is encouraging. Several formulations of the fungus Trichoderma harzianium are now come to market as a control for grey mold (Botrytis). Other biofungicides now available include a control for powdery mildew in grapes and a protectant against tree wound pathogens.

References

- Organic-farming, crop-production, agriculture: vikaspedia.in, Retrieved 4 June, 2019

- Organic-farming-benefits: conserve-energy-future.com, Retrieved 18 January, 2019

- What-is-organic-farming: organicconsumers.org, Retrieved 29 August, 2019

- Coleman, Eliot (1995), The New Organic Grower: A Master's Manual of Tools and Techniques for the Home and Market Gardener (2nd ed.), pp. 65, 108, ISBN 978-0930031756

- Organic-gardening: maximumyield.com, Retrieved 9 July, 2019

- Horne, Paul Anthony (2008). Integrated pest management for crops and pastures. CSIRO Publishing. p. 2. ISBN 978-0-643-09257-0

- Organic-garden: conserve-energy-future.com, Retrieved 11 April, 2019

- Organic-food: britannica.com, Retrieved 13 February, 2019

- Blair, Robert. (2012). Organic Production and Food Quality: A Down to Earth Analysis. Wiley-Blackwell, Oxford, UK. Pages 72, 223. ISBN 978-0-8138-1217-5

- Organic-foods, healthy-eating: helpguide.org, Retrieved 28 March, 2019

- Organic-vegetable-production: extension.psu.edu, Retrieved 8 May, 2019

Permissions

Index

www.ingramcontent.com/pod-product-compliance
Lightning Source LLC
Chambersburg PA
CBHW061253190326
41458CB00011B/3661